Alternative Transportation Fuels and Vehicle Technologies

Challenges and Opportunities

A Report of the CSIS Energy and National Security Program,
CSIS Global Strategy Institute, and National Renewable Energy Laboratory

AUTHORS
Douglas Arent
Frank A. Verrastro
Jennifer L. Bovair
Erik R. Peterson

FOREWORD
Senator Richard G. Lugar

CONTRIBUTORS
Sarah O. Ladislaw
David Pumphrey

August 2009

CSIS | CENTER FOR STRATEGIC &
INTERNATIONAL STUDIES

About CSIS

In an era of ever-changing global opportunities and challenges, the Center for Strategic and International Studies (CSIS) provides strategic insights and practical policy solutions to decisionmakers. CSIS conducts research and analysis and develops policy initiatives that look into the future and anticipate change.

Founded by David M. Abshire and Admiral Arleigh Burke at the height of the Cold War, CSIS was dedicated to the simple but urgent goal of finding ways for America to survive as a nation and prosper as a people. Since 1962, CSIS has grown to become one of the world's preeminent public policy institutions.

Today, CSIS is a bipartisan, nonprofit organization headquartered in Washington, D.C. More than 220 full-time staff and a large network of affiliated scholars focus their expertise on defense and security; on the world's regions and the unique challenges inherent to them; and on the issues that know no boundary in an increasingly connected world.

Former U.S. senator Sam Nunn became chairman of the CSIS Board of Trustees in 1999, and John J. Hamre has led CSIS as its president and chief executive officer since 2000.

CSIS does not take specific policy positions; accordingly, all views expressed in this publication should be understood to be solely those of the author(s).

About NREL

The National Renewable Energy Laboratory (NREL) is the nation's primary laboratory for renewable energy and energy efficiency research and development. NREL's mission is to advance the Department of Energy's (DOE) and our nation's energy goals. Scientific research focuses on the understanding renewable resources for energy, conversion of these resources to renewable electricity and fuels, and the use of renewable electricity and fuels in homes, commercial buildings, and vehicles. NREL's R&D areas of expertise are renewable electricity, renewable fuels, integrated energy system engineering and testing, strategic energy analysis, and technology transfer. It is home to the National Center for Photovoltaics, the National Bioenergy Center, and the National Wind Technology Center.

Library of Congress Cataloguing-in-Publication Data
CIP information available on request.
ISBN 978-0-89206-542-4

The CSIS Press
Center for Strategic and International Studies
1800 K Street, N.W., Washington, D.C. 20006
Tel: (202) 775-3119
Fax: (202) 775-3199
Web: www.csis.org

CONTENTS

ACKNOWLEDGMENTS

The authors gratefully acknowledge the help of their colleagues throughout the production of this report. For research and other support, our gratitude goes out to Kartik Akileswaran, J. Phillip Behm, Tiffany Blanchard, James Coan, Matthew Frank, Bill Gingher, Brendan Harney, Alexander Iannaccone, Dae Woo Lee, Catherine Mason, Molly Middaugh, Zack Poindexter, Rachel Posner, Brian Stevens, John Tincoff, Molly Walton, and Sarah Yates. Any errors or omissions are the responsibility of the authors.

The *Alternative Transportation Fuels and Vehicle Technologies* report and series is made possible through the generous support of the Aramco Services Company, BP, and Chevron. The authors acknowledge the assistance of the following institutions in the preparation of this report and the attendant series from which this summary is drawn: U.S. Department of Energy, National Renewable Energy Laboratory, Association of Oil Pipelines, University of California–Berkeley, Georgetown University, Massachusetts Institute of Technology, Iowa State University, Shell Oil, Dupont, Energy Future Coalition, Princeton Environmental Institute, Argonne National Laboratory, Conoco Phillips, Syntroleum, General Motors Corp., Toyota Motors, National Petrochemical and Refiners Association, R.L. Banks and Associates, Inc., and the Rocky Mountain Institute.

FOREWORD
Senator Richard G. Lugar

In a speech I gave in March of 2006, I asserted that exploding demand for energy, the vulnerability of energy supplies to terrorism and warfare, the increasing concentration of energy assets in the hands of problematic governments, the growing willingness of these governments to use energy as a geopolitical weapon, and evidence that climate change was accelerating had combined to change the energy debate fundamentally. I contended that the "balance of realism" in energy policy had shifted from proponents of a laissez-faire approach that would continue U.S. overdependence on oil to advocates of energy alternatives who recognize the urgency of achieving a major change in the way that the United States and the world produces and uses energy.

In the intervening years, public awareness of our energy dilemma has greatly improved. Politicians understand that Americans care about issues like energy security and the need to address environmental concerns. Yet, despite this growing focus and rising prices, we have yet to commit ourselves fully to the policy steps required to achieve a sustainable future.

The world needs alternatives to oil. This requires a reorientation of our transportation sector and a more diverse set of options for fuel choices. The CSIS Energy and National Security Program and Global Strategy Institute and the National Renewable Energy Lab (NREL) have for years provided vital policy analysis, guidance, and world-class research on energy-related topics. I applaud them for collaborating to produce this valuable report. It draws needed attention to the challenges we face in transportation fuel supply and examines opportunities for solutions.

The global oil market has fundamentally shifted under pressure from surging demand and tightening supply margins. Spare capacity has shrunk from as much as 10 percent just five years ago to around 3 percent. This means that relatively small oil supply losses can have dramatic effects on world prices. Small margins also make political manipulation of supplies a more potent weapon against the United States and import-dependent countries. As competition for scarce oil resources grows and the price of oil stays high, oil will be an even greater magnet for conflict.

In my service in the U.S. Senate on the Foreign Relations Committee, I have seen the many ways in which energy constrains our foreign policy options, limiting effectiveness in some cases and forcing our hand in others. We pressure Sudan to stop genocide in Darfur, but we find that the Sudanese government is insulated by oil revenue and oil supply relationships. We pressure Iran to stop its uranium enrichment activities, yet key nations are hesitant to support us for fear of endangering their access to Iran's oil and natural gas. We try to foster global respect for civil society and human rights, yet oil revenues flowing to authoritarian governments are often diverted to corrupt or repressive purposes. We fight terrorism, yet some of the hundreds of billions of dollars we spend each year on oil imports is diverted to terrorists. We give foreign assistance to lift people out of poverty, yet energy-poor countries are further impoverished by expensive energy-import bills. We seek options that would allow for military disengagement in Iraq and the wider Middle East, yet our way of life depends on a steady stream of oil from that region.

Ending our oil import dependence will not suddenly cure poverty, end terrorism, prevent weapons proliferation, or bring peace to the Mideast. But failing to address energy guarantees that our pursuit of these foreign policy goals will be encumbered and our way of life will remain under threat. American national security will be at risk as long as we are excessively dependent on imported oil.

Recently, the International Energy Agency (IEA) projected that global energy demand would increase by 45 percent by 2030. Three-quarters of this demand growth is likely to happen in the developing world, with 43 percent of it happening in China and India alone. Eighty-four percent of that demand growth is expected to come from fossil fuels, translating into nearly a 50 percent increase in carbon dioxide emissions. We can debate the margin of error in any of these international energy projections, but the picture they paint is a bleak one for global stability and U.S. influence.

Awareness of our energy dilemma is improving. Yet, advancements in American energy security have been painfully slow, and political leadership has been defensive, rather than proactive. If we suffer an oil embargo, if terrorists succeed in disrupting our oil lifeline, if we slide into a military conflict because oil wealth has emboldened anti-American regimes, or if climate change is accelerated by unrestrained growth in carbon emissions, it will not matter that before disaster struck, the American public and its leaders gained a new sense of realism about our vulnerability. We need to have the discipline to understand that a modestly positive trend line in replacing oil is not enough.

Technological breakthroughs that expand energy supplies for billions of people worldwide will be necessary for sustained economic growth. If concerns over climate change are factored into policies, the challenge becomes even greater. The successful development and deployment of new energy technologies face hurdles well beyond price. Our nation's transportation infrastructure has been built over a period of a hundred years around the premise of cheap and accessible oil—a premise that no longer holds. Americans are not free to choose transportation fuels other than those based on petroleum, and we have limited vehicle choices as well. There are substantial barriers to moving new technologies to market—from inconsistent R&D funding, to logistical constraints, to the complex interface between energy, food, economic growth, and demographic change.

We can overcome these challenges. The United States has the financial resources, scientific prowess, productive land, and industrial infrastructure to address our energy vulnerability. The question is whether we will heed the abundant warning signs and apply the leadership and political will to deal with this problem in the present rather than suffering grave consequences in the future.

I congratulate CSIS and NREL for promoting our vital interest in finding alternatives to oil and improving policy discussion on possible solutions.

—Richard G. Lugar, U.S. Senator

1 | INTRODUCTION

The Looming Energy Challenge

The global energy system is changing. New demand centers are emerging, and there are numerous challenges to expanding the transportation fuels infrastructure. It has become clear that without significant changes in policy or the introduction of new technologies, the world will continue on an unsustainable path with respect to how it produces, delivers, and uses its energy resources. The confluence of heightened concerns over U.S. energy security, volatile swings in the price of oil, and awareness of climate change has refocused efforts to reduce petroleum demand, improve efficiency, and spur development of zero or low emissions fuels.

In the United States, oil use accounts for 40 percent of total energy consumption and almost three-quarters of petroleum use is dedicated to meeting transport needs.[1] More than 95 percent of current transport fuels are derived from petroleum.[2] In 2007, the United States consumed more than 20 million barrels per day (b/d) of petroleum products. Absent significant changes in automotive technologies, fuel efficiency, driving patterns, and fuel choices, oil is projected to dominate as the primary transportation fuel in the United States. Although a few years of high prices and a global economic downturn have resulted in lower oil demand and a precipitous drop in prices, business as usual (BAU) projections forecast a growing appetite for oil. Even with a flattening of demand, some analysts have predicted that by 2030, the United States will need to import about 17 million b/d, roughly three-fourths of which will be used to meet transportation demand.[3]

At the same time as the U.S. thirst for oil continues, total global liquids consumption is projected to grow by more than 25 percent by 2030—from nearly 84 million b/d in 2007 to more than 106 million b/d.[4] The scale and rapidity of this level of growth are unprecedented. While it took over a century to consume the first trillion barrels of oil, the second trillion will be gone in a few decades. Consequently, even without the overlay of prevailing climate concerns, there are a host of compelling reasons for a closer examination of fuel options, enabling technologies, and the efficiency of the transportation sector.

---

1. Energy Information Administration (EIA), *Annual Energy Outlook 2008* (Washington, D.C.: U.S. Department of Energy, June 2008).

2. Bureau of Transportation Statistics, *National Transportation Statistics 2007* (Washington, D.C.: U.S. Department of Transportation, 2007), 261, table 4-2, http://www.bts.gov/publications/national_transportation_statistics/html/table_04_02.html.

3. EIA, *Short-Term Energy Outlook* (Washington, D.C.: U.S. Department of Energy, July 8, 2008), http://www.eia.doe.gov/emeu/steo/pub/jul08.pdf.

4. EIA, *International Energy Outlook 2008* (Washington, D.C.: U.S. Department of Energy, June 2008), http://www.eia.doc.gov/oiaf/ieo/liquid_fuels.html.

Figure 1. World Liquids Consumption by Sector, 2004–2030[5]

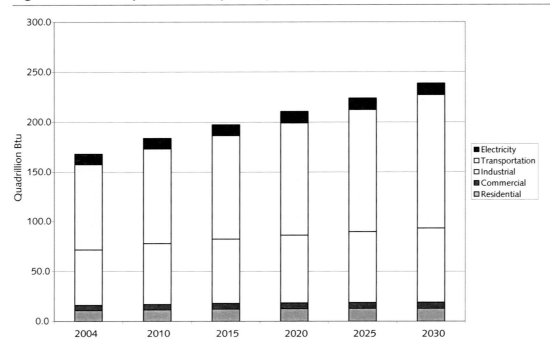

The reality of American life is that transportation plays a key role in the economy, and economic growth has been facilitated by low energy prices. Historically, individuals and businesses have relied on fossil fuels to meet their energy needs: from commuting to work to shipping products to keeping food on the table. The sharp rise in gasoline prices from 2005 to 2008 spurred both a reminder of oil shocks of the past and new fears about the future. A convergence of factors brought about record-level oil prices far above previous projections. The market fundamentals of supply and demand in an era of rapid global demand growth, geopolitical tensions, tightened refinery capacity, lack of infrastructure investment, as well as the new role of commodity investors contributed to the price trajectory and sharp climb in the price of oil. As price increased, consumption patterns and consumer behavior began to adapt to new realities; however, it took a global economic meltdown coupled with a credit crisis to cause oil prices to recede to current levels. Concerns remain about the long-term affordability, availability, and environmental impacts of the continued use of the internal combustion engine and the current array of transportation fuels.

There are many actions that both private markets and U.S. policymakers can take to alter the energy demand curve for future growth rates and moderate or potentially reverse the trend of the growing demand for oil imports. Several of these actions will be discussed in detail in this report, including policies aimed at diversifying transportation fuel options, improving the fuel delivery infrastructure, and promoting improvements in automotive efficiency and fuel conservation. Each of these actions has various costs, benefits, and policy trade-offs. Owing to the scale of our transport system, implementation of any of these approaches will require significant lead time in order to make a noticeable impact on consumption and behavior.

5. EIA, *International Energy Outlook 2008*, chapter 2, "liquid fuels," figure 28, http://www.eia.doc.gov/oiaf/ieo/liquid_fuels.html.

In addition, heightened awareness of climate change is changing the terms of the U.S. energy equation. Decisions on climate change policy are a major "game-changing" factor. This is a new challenge that energy producers and consumers alike must face. The realistic prospect of a global market (price/tax/cap) for carbon will increase both the technical complexity and the cost of using fossil fuels for transportation (as well as other sectors that are not the subject of this brief). Increased efficiency, carbon mitigation strategies, and lower-carbon (zero or low emissions) fuels will all need to be further explored, developed, and integrated into existing energy markets in order to meet this challenge.

Taken together, the economic, security, and environmental challenges that loom in the U.S. energy future have renewed interest in alternative fuels and vehicle technologies. To address this issue, the CSIS Energy and National Security Program and Global Strategy Institute, together with the National Renewable Energy Laboratory (NREL), hosted a series of conferences on the production, conversion, delivery, and consumption of energy in the transportation sector. The purpose of the series was to bring together academics, researchers, energy producers, transporters, manufacturers, consumers, and policymakers to examine two major components of the energy challenge facing the United States in coming decades: the range of fuel choices that might be made available and the introduction of new automotive technologies.

The alternative fuels examined over the course of the series included liquid substitutes for petroleum, such as ethanol, biobutanol, renewable diesel and biodiesel, gas-to-liquids, coal-to-liquids, and other carbon-to-liquids options. Vehicle technologies included advanced internal combustion engines, hybrid electric, plug-in hybrids, and flex-fuel vehicles. Opportunities for emerging technologies such as fuel cells were also examined. As the number of alternative fuels and vehicles increases to meet growing demand and to supplement petroleum-based gasoline, the transportation sector will become more diverse, and the delivery infrastructure more complex and perhaps less fungible. The purpose of the series was to look at all of the options on the table without advocating for specific fuels or attempting to pick technology winners.

While the findings of the series clearly showed that the notion of energy independence should be replaced with synergistic energy interdependence, many of the alternative fuels and technologies available for transportation use can be indigenously sourced and developed. In addition to helping meet energy demand, alternatives would also serve to lessen geopolitical and environmental risks associated with continued conventional oil dependence. The further expansion of these alternatives can place the United States in the position of leveraged interdependence. In this case, the United States could theoretically move from the position of a price taker that is reliant on imports of a critical resource to a position where it would potentially enjoy some measure of increased advantage and stability. Development of alternatives could also help level the playing field between nations that are current importers of oil and those that are conventional producers. The combined use of some new low emission alternatives and increased vehicle efficiency would also move the United States on a path toward reduced carbon emissions, greater environmental sustainability, and enhanced energy security.

This report attempts to present a side-by-side comparison of fuels and vehicles with the recognition that future technological advances or additional research into unresolved issues, such as the land-use impact of biofuel production, could alter findings and make certain options more or less attractive. The report examines the benefits and challenges of various alternatives, including resource availability, infrastructure needs, capital requirements, timetables for at-scale contributions, and environmental sustainability. Chapter 2 looks at new alternative fuels, from hydrocarbon products to biofuels, and provides an overview of the infrastructure challenge the United States faces when looking to implement any of these new alternatives on a mass scale. Chapter 3 focuses on advances in technologies for light-duty vehicles including high-efficiency internal-combustion engines, flex-fuel vehicles, and hybrid and electric cars. Chapter 4 addresses the challenges in adopting new policies and creating incentives for change.

2 FUEL CHOICES

Unconventional Fossil Fuels

Projections from both of the leading energy statistical agencies, the Energy Information Administration (EIA) and the International Energy Agency (IEA), predict that fossil fuels, oil in particular, will continue to be the most widely used transportation fuel sources for decades to come. While additional liquid fuel options for transportation and energy are expanding, enacting policies to address new legislative goals will be particularly challenging, especially when one considers the size of the U.S. transportation fleet and consumption growth forecasts. Consequently, we have to be both visionary and at the same time thoughtful as we establish expectations for transforming the transportation fuel sector.

Recent geological studies have shown that the resource endowment—that is, "molecules in the ground"—is, indeed, enormous.[1] While there are significant environmental and economic challenges associated with full-scale development, the extent of the total hydrocarbon resource base in the United States remains quite vast—large enough to equal seven times current world proven oil reserves, as shown in Figure 2.

If the United States were able to convert all estimated hydrocarbon resources into liquid fuel for transportation purposes, domestic fossil fuel resources would be enough to supply future domestic demand, even at its highest estimates, for a century or more. Harnessing even a small percentage of the total energy potential stored in domestic resources would increase U.S. production, help to meet future demand growth, and decrease dependence on imported oil.[2] The "above ground" challenges to doing so, however, are many and varied, encompassing an array of technical, political, environmental, and economic (including investment and infrastructure) considerations.

When looking ahead at a future of volatile energy prices and new technological advances, there are several potential unconventional hydrocarbon options. Five main alternatives exist: oil shale, oil (tar) sands, natural gas-to-liquids, coal-to-liquids, and methane hydrates. When considering time, scale, cost, and technology, hydrates development is seen as a longer-term option. Consequently, the first four alternatives are discussed here in detail.

---

1. National Petroleum Council (NPC), *Hard Truths: Facing the Hard Truths about Energy* (Washington, D.C.: NPC, July 2007).

2. Reducing energy dependence will improve energy security by reducing the prospects of energy shortages and wealth transfer to nations abroad. However, no nation is entirely self-sufficient in terms of energy resources, and it will be extremely costly to actually achieve a goal of total energy independence. Thus, while reducing dependence is a laudable directional goal, the notion of total energy independence is unreasonable and fails to account for the global nature of the energy market and the various tradeoffs one encounters when formulating energy policy.

Figure 2. U.S. Hydrocarbon Resources[3]

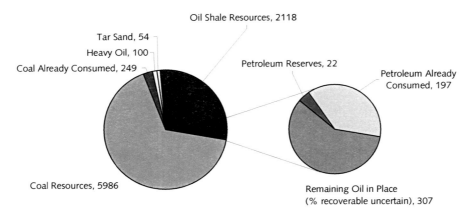

Oil Shale

Oil shale is a sedimentary rock containing oil substances in solid form. To separate and extract oil from the rock formations, oil shale must be mined and heated in a process known as retorting. Oil shale is a very dense fuel in terms of energy concentration; one ton of oil shale can yield approximately 25 gallons of oil.[4] The United States contains the world's largest oil shale deposits, accounting for nearly 2.1 trillion barrels. The Green River Formation, covering parts of Wyoming, Utah, and Colorado, is estimated to hold 1.5–1.8 trillion potentially recoverable barrels.[5] Many countries currently maintain, or have pursued, programs to develop oil shale deposits, but even today the resource remains relatively unexploited. Only Brazil, China, and Estonia produced oil through the processing of oil shale as of 2005, with a collective output of 13.6 thousand b/d.[6]

Environmental challenges pose the primary impediment to further development and use of oil shale. Mining and extraction operations are ecologically intensive, raising land disturbance and reclamation issues.[7] This issue must be addressed before commercial-scale oil shale processing will move forward in the United States.

Shell Oil has developed a process known as "in-situ conversion," which is believed be lower cost and less detrimental to the environment than current mining and refining processes. During in-situ conversion, oil shale is slowly heated underground over a period of a few years to 700°F, separating the oil, which is then pumped through a series of wells to the surface. This process is claimed to be economically feasible with crude oil prices in excess of $30 per barrel, although Shell has yet to expand production to commercial scale. Shell is currently evaluating the use of renewable electricity generation to power the in-situ conversion to reduce the overall environmental impact of the fuel.

3. David Garman, "Addressing America's Petroleum Independence" (presentation, CSIS, Washington, D.C., June 15, 2006), http://www.csis.org/media/csis/events/060615_garman.pdf.

4. Harry R. Johnson et al., *Strategic Significance of America's Oil Shale Resource: Volume 1: Assessment of Strategic Issues* (Washington, D.C.: U.S. Department of Energy, March 2004).

5. James T. Bartis et al., *Oil Shale Development in the United States: Prospects and Policy Issues* (Santa Monica, C.A.: RAND, 2005).

6. World Energy Council, *2007 Survey of Energy Resources* (London: World Energy Council, 2007), 93–119.

7. Office of Petroleum Reserves, U.S. Department of Energy, "Fact Sheet: Oil Shale Water Resources," http://www.fossil.energy.gov/programs/reserves/npr/Oil_Shale_Water_Requirements.pdf.

Oil (Tar) Sands

Oil sands are a naturally occurring viscous mixture of sand or clay, water, and an extra heavy substance called bitumen. Unless heated or diluted, bitumen will not flow. Where oil sands are located near the surface, extraction occurs through open-pit mining, after which the sand is mixed with warm water to separate the bitumen. For accessing oil sands farther below the surface, newer extraction technologies involve variations of "in-situ" steam injection, which allow the bitumen to be heated in place underground and then pumped by well to the surface. Because its viscosity is much greater than that of conventional crude oil, bitumen must be refined as synthetic crude through petroleum coking or hydrocracking and reforming. Roughly 1.16 barrels of extracted bitumen are required to produce 1 barrel of synthetic crude.[8]

Oil and tar sands resource deposits are concentrated overwhelmingly in western Canada and Venezuela, estimated to hold 2.5 trillion and 1.5 trillion barrels of oil equivalent (BOE) in place, respectively. Economically recoverable reserves are projected to be 175 billion BOE from Canada's Albertan Oil Sands and 270 billion BOE from Venezuela's Orinoco Belt. In the United States, the most significant oil sands deposits are in Utah, estimated to contain 32 billion BOE. Commercial-scale oil sand extraction and conversion technology has been utilized in Canada for some time, where oil sands production accounted for more than a third[9] of Canada's 3.29 million b/d of oil output in 2006.[10] Production is claimed to be economically viable with synthetic crude oil prices in excess of $30 per barrel,[11] but increasingly there have been concerns about generally rising operational costs and the availability of skilled labor to meet more ambitious production targets, as well as potentially severe environmental impacts.[12]

Liquid fuel production from oil sands requires a large input of both natural gas and large volumes of water. While recent drilling success has produced an increase in U.S. domestic gas production volumes (reversing earlier declines), this is due in large part to increased output from nonconventional (shale) gas resources. The boom in shale gas development in 2007 and 2008 was largely driven by access, technology, and higher prices. And while potentially large resources remain undeveloped at this writing, it is unclear how the downturn in price and the financial crisis will affect future output. As an alternative to gas, coal and nuclear energy are being considered, each with its own set of environmental and other concerns.

Natural Gas-to-Liquids

Gas-to-liquid (GTL) fuel utilizes Fischer-Tropsch (FT) technology to convert natural gas into usable liquid fuel for diesel engine applications in cars, buses, jets, and heavy equipment.[13] GTL proj-

---

8. Energy Information Administration (EIA), *Annual Energy Outlook 2006* (Washington, D.C.: U.S. Department of Energy, 2006).

9. Alberta Department of Energy, "Alberta's Oil Sands 2006," December 27, 2007, http://www.energy. gov.ab.ca/OilSands/pdfs/Osgenbrf.pdf.

10. EIA, "Country Energy Profiles," Top World Oil Producers and Consumers, 2006, http://www.eia. doe.gov/emeu/cabs/topworldtables1_2.htm.

11. EIA, *Annual Energy Outlook 2006*.

12. Doug Struck, "Dollars and Doldrums Mix in Canada's Oil Boomtown," *Washington Post*, January 1, 2007.

13. Security issues may arise from using natural gas for GTL if it leads to or exacerbates the need for North America to increase imports of liquefied natural gas (LNG). The largest reserves of natural gas are found in Iran and Russia.

ects target smaller gas reserves (e.g., the 1,300-plus known gas fields that are estimated to contain between 0.5 and 5 trillion cubic feet of natural gas resources). Because of their size, these smaller accumulations are generally uneconomic to be developed as liquid natural gas products.

Shell, Syntroleum, and Sasol are among the handful of companies that have developed several variations of GTL Fischer-Tropsch technology. These companies currently operate pilot plants supplying a limited amount of GTL fuel to vehicle test fleets for demonstration purposes. Existing production capacity is close to 50,000 b/d and expected to increase to nearly 500,000 b/d in 2030 if currently planned projects are completed.[14] Commercial-scale GTL production facilities are being built in Nigeria and Qatar, but significant cost overruns have delayed completion of those projects and resulted in the cancellation of others. In 2006, the EIA estimated capital costs of building GTL plants to be $25,000 to $45,000 per barrel of daily capacity. In terms of economic viability, the Department of Energy provides a price range for oil needed in order to have production of GTL. Assuming Henry Hub natural gas prices of between $5.82 and $7.22 per million Btu (2006 dollars; $5.99–$7.43 per thousand cubic feet) between now and 2030, it expects no GTL production in a reference oil price scenario of between $57 and $70 per bbl (2006 dollars).[15] Yet in a high price oil case that steadily rises to $119/bbl in 2030, it expects GTL production of about 250,000 b/d in 2030.

Coal-to-Liquids

Coal-to-liquid (CTL) fuel production, like GTL, utilizes Fischer-Tropsch technology, where coal is processed into liquid diesel fuel by one of two methods: either indirect coal liquefaction (ICL) or direct coal liquefaction (DCL). In ICL, the coal is first gasified, and then synthetic fuel is made from the gas. In DCL, the coal is heated, pressurized, and then processed with a catalyst. DCL, the preferred conversion process, has higher conversion efficiency, although several technical challenges remain before the process can be developed at scale.

At this time, Sasol is the only commercial producer of CTL fuel in the world, operating a 150,000 b/d facility in South Africa. In China, China Shenhua Coal Liquefaction Corp. Ltd. is building two 80,000 b/d CTL plants, which are expected to begin production early in the next decade, each at a cost of $6 billion to $7 billion.[16] Although there has been much recent enthusiasm, CTL technology exists in the United States only in laboratories and at pilot plants, due to the many economic, technical, and environmental barriers that first must be overcome before it can be expanded at scale. Despite these challenges, the EIA estimates U.S. production of CTL fuel to increase to 400,000 barrels of oil equivalent by 2030.[17]

---

14. Andrew J. Slaughter et al., "Topic Paper #9: Gas to Liquids (GTL)," Working Document of the National Petroleum Council (NPC) Global Oil and Gas Study, Washington, D.C., July 18, 2007, http://www.npchardtruthsreport.org/topic_papers.php.

15. EIA, *Annual Energy Outlook 2008* (Washington, D.C.: U.S. Department of Energy, 2008).

16. EIA, *Annual Energy Outlook 2006*.

17. EIA, *Annual Energy Outlook 2007* (Washington, D.C.: U.S. Department of Energy, 2007).

Figure 3. Coal-to-Fuels Production Cycle[18]

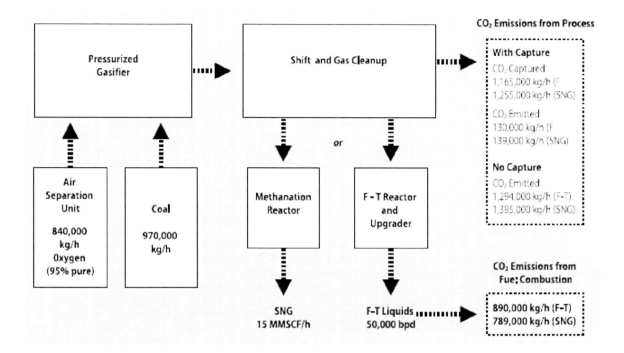

The relatively low crude oil price of the past few decades halted domestic expansion of CTL technology. Costs of capital, financing, and operation, as well as upcoming regulation (including carbon emission costs from fuel and operations), will determine the economic viability of a CTL plant. Estimates as late as 2006, with crude oil above $40/barrel,[19] suggested possible profitability for full-scale plants. Deteriorating economic conditions, combined with volatile crude oil prices and likely climate legislation, however, have deterred technology development and investment at present.[20]

Despite the domestic availability of coal, CTL fuel production faces a host of environmental challenges that have impeded and will continue to challenge large-scale expansion of CTL technology. There is strong concern regarding waste and wastewater treatment and disposal. Perhaps the greatest hurdle to overcome, however, is the high carbon emissions intensity associated with CTL fuel production. Even if it were technically possible to capture 90 percent of the carbon emissions from CTL processes, the total carbon emitted would still be higher than that associated with traditional petroleum production; accounting for the additional carbon released when CTL fuel is burned in an internal combustion engine, the well-to-wheels carbon emissions from CTL in combination with a carbon capture system, is still 8 percent higher than for petroleum.[21]

---

18. John Deutch et al., *The Future of Coal: Options for a Carbon-Constrained World* (Cambridge, MA: MIT, 2007), 155.

19. EIA, *Annual Energy Outlook 2006*.

20. Office of Air and Radiation, "Technical Support Document: Coal-to-Liquids Product Industry Overview," Environmental Protection Agency (EPA), Washington, D.C., January 28, 2009, www.epa.gov/climatechange/emissions/downloads/tsd/TSD%20CTL%20suppliers_013009.pdf.

21. Natural Resources Defense Council, "Climate Facts: Why Liquid Coal is Not a Viable Option to Move America Beyond Oil," February 2007, http://www.nrdc.org/globalWarming/coal/liquids.pdf.

Biofuels

Only recently has there been sustained interest in biofuels as a replacement or supplement to traditional petroleum-based transportation fuels in the United States. Years of record-high oil prices in combination with concerns over carbon emissions and energy security spurred rapid worldwide development of biofuels production capacity. This enthusiasm, however, has been tempered with the realities of introducing biofuels at a scale of more than a few percent of domestic consumption, particularly for current commercial pathways of corn, sugar cane–based ethanol, and soy and palm–based biodiesel. This is due to numerous commercial reasons, as well as increasing scrutiny of total environmental impacts of these specific feedstock/pathway options, and more recently, concerns about the impact of biofuels production on grain and food availability and prices.

While cellulosic ethanol has been the focus of considerable research and development, scale-up engineering, and investment requirements, the technologies and processes for conversion of cellulose to biofuels have not yet made production from these materials economically attractive. Before biofuels can become a more substantial part of the transportation fuels mix, there are infrastructure challenges, capital requirements, and trade-offs with land, water, and food resources. As bio-based fuels continue to expand, production technologies that leverage the enormous existing fuel refining and distribution infrastructure will have significant advantages over biofuels that require new production facilities and independent distribution and blending infrastructure.

The biofuels category is far broader than ethanol and includes products such as biobutanol, biodiesel, and renewable diesel, as well as other possible molecules derived from bioresources. Inputs for conversion to biofuels can be sourced from a variety of renewable feedstocks including sugar cane, corn, and soybeans; waste streams, grasses, and wood plants; and byproducts of agricultural and forestry production. Conversion technologies enable molecular readjustment of biomass, breaking down feedstock materials and reassembling the molecules to form liquid fuels that are compatible with existing engines.

Both fossil energy and biomass resources can be converted into liquid transportation fuels via a number of processes. Figure 4 illustrates the main routes to produce transportation fuels from fossil fuels and biomass. Biomass resources fall fundamentally into four categories: sugar/starch, lignocelluloses or the fibrous components, oil crops (such as soy, palm, rape seed, and jatropha oils), and hydrous/wet biomass. For the latter, anaerobic gasification is a commercially proven technology and used widely around the world. The primary output is methane (i.e., natural gas), which is most frequently used directly for electricity generation or process heat. Alternatively, gasification can be used to produce syngas that can be converted to methanol, Fischer-Tropsch liquids, dimethyl ether (DME), and hydrogen. Hydrothermal treatment—an alternative conversion process still under development—can be used to produce "biocrude" that can be further processed into middle distillates, such as diesel.

Figure 4. Pathways to Liquid Fuels[22]

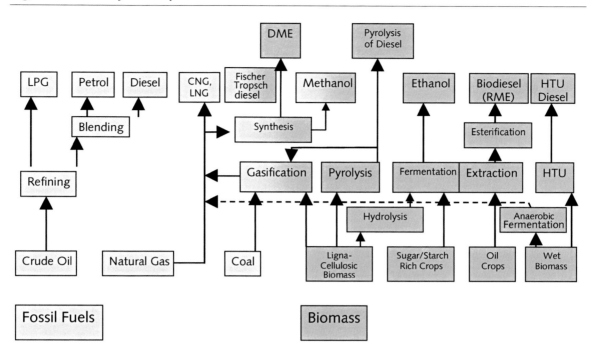

Oil crops (such as soy, palm, rapeseed, jatropha oils) or waste grease or fats form the basis of bio-based middle distillates (e.g. diesel), which are commercially produced via a number of technologies. Biodiesel, a fatty acid methyl ester (a different chemical compound than fossil fuel–derived diesel, which does not contain the "ester" unit), is predominantly produced through a process called esterification. This process is relatively straightforward, involving the addition of acids (such as sulphuric acid) or bases (such as potassium hydroxide) to produce a biocrude that is further refined. Glycerine is the primary by-product of this pathway. Alternatively, these oils can be converted by gasification or by pyrolysis into a biocrude for further refining.

Production of gasoline-compatible bio-based fuels such as ethanol and butanol can take place via direct fermentation of sugar- and starch-rich biomass, the most utilized route for production of ethanol to date, or this can be preceded by hydrolysis processes to convert lignocellulosic biomass (e.g., the woody fibrous materials) to sugars first. A wide variety of feedstock and conversion processes for bioethanol or other light distillates exists and is the subject of continuing technical research and political debate.

Clearly, the breadth of processing pathways, as well as fossil fuel resources, suggests that alternative fuels policies and approaches should not be restricted to only a few. That is, significant opportunity exists for technical and economic efficiency while still adhering to the core focus of an alternative fuels strategy: economic productivity, enhanced energy security, and environmental stewardship. When we look ahead, the main driving factors behind the use of liquid transportation fuels are identifying and producing the supplies to meet growing demand; environmentally sound extraction and conversion methods; economics; investment issues, including delivery infrastructure; and considerations related to greenhouse gas emissions and carbon constraints.

22. Adapted from E. Van Thuijl et al., *An Overview of Biofuel Technologies, Markets and Policies in Europe* (Amsterdam: Energy Research Center of the Netherlands, January 2003).

Two principal pathways exist for conversion of bioresources to fuels: thermochemical and biochemical. As previously described, within each, there are multiple process options, as show in Figure 5. Gasification can be used to produce multiple products. Production of ethanol (or butanol) can take place via direct fermentation or via conversion of the lingo-cellulosic components. The most predominant pathways are illustrated in greater detail.

Figure 5. Main Conversion Options for Biomass to Secondary Energy Carriers[23]

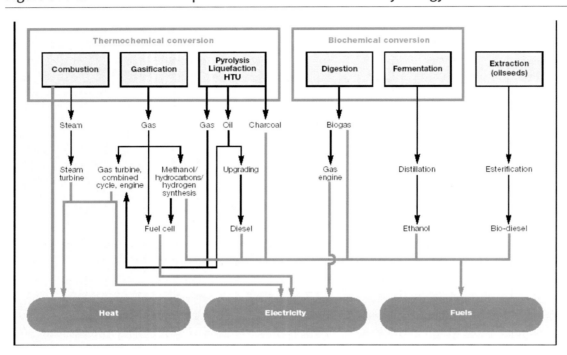

The characteristics of fuels differ widely (see Table 1). All fuels considered, except DME, are liquids and can be stored and distributed with relatively conventional infrastructure; however, infrastructure considerations are a major issue discussed in further detail below. Ethanol is particularly problematic for multiuse pipelines due to its hydrophilic properties (i.e., it mixes easily with water). Although tests are currently underway to explore options for pipeline batching and transport, to date the challenge requires independent transportation, distribution, and storage before blending. Other concerns include the range of toxicity and environmental impacts (which is particularly dependent on land-use impacts, farming, and production processes) of some of the fuels.[24] The state of technology pathways for the manufacture of biofuels is summarized below, recognizing that there is a plethora of information available.

---

23. Wim C. Turkenburg, "Renewable Energy Technologies," in *World Energy Assessment: Energy and the Challenge of Sustainability*, ed. Jose Goldemberg (New York: United Nations Development Programme/UN-DESA /World Energy Council, 2000), http://www.energyandenvironment.undp.org/undp/indexAction.cfm ?module=Library&action=GetFile&DocumentAttachmentID=1020. Some categories represent a wide range of technological concepts as well as capacity ranges at which they are deployed, which are dealt with further in the main text.

24. André Faaij, "Modern Biomass Conversion Technologies," *Mitigation and Adaptation Strategies for Global Change* 11, no. 2 (March 2006): 335–367.

While there are clearly some distinct advantages associated with biofuels, water requirements are typically greater than those needed for conventional oil production. However, assuming no land use impacts, research indicates that biofuels generally emit less CO_2 per unit energy as compared with conventional oil. The reduction is relatively small for corn ethanol and soybean biodiesel, about 15 to 20 percent lower, although the range is larger depending on assumptions. Much larger reductions of 80 percent or greater are calculated for cellulosic and sugar cane biofuels. With land use displacement, though, recent articles argue that even cellulosic and sugar cane biofuels may emit more CO_2 than conventional oil.[25]

Table 1: Fuel Characteristics

Fuel	Energy Content (BTU/gallon)	Other aspects
Methanol	66,000	Toxic in direct contact Octane number 88.6 (gasoline 85)
DME	67,000	Gaseous Vapour pressure 5.1 bar at 20°C
Fischer-Tropsch gasoline	127,000	Very comparable to diesel and gasoline; zero sulfur, no aromates
Ethanol	84,000	Octane number 89.7 (gasoline 85)
Diesel from bio-oil/ bio-crude	144,000	When fully de-oxygenated
Biodiesel	133,000	Cetane number: 58 (diesel 47.5)
Gasoline	115,000	Depending on refining process, contains sulphur and aromates
Diesel	139,000	Depending on refining process, contains sulphur and aromates

Ethanol

In current U.S. commercial practice, starchy sugars of corn are harnessed through biochemical conversion processes of fermentation. The United States produced 9.3 billion gallons of ethanol in 2008.[26] The majority of U.S.-produced ethanol is derived from corn at more than 100 refineries.[27] Moreover, while that volume is expected to increase, it should be noted that in terms of energy content (rather than simple volumetric comparisons) 1 billion gallons per year of corn ethanol production roughly equates to about 45 thousand barrels per day of gasoline displacement.

---

25. See Timothy Searchinger et al., "Use of U.S. Croplands for Biofuels Increases Greenhouse Gases through Emissions from Land-Use Change," *Science* 319, no. 5867 (February 29, 2008): 1238–1240.

26. U.S. Grains Council's International Distillers Grains Conference, Indianapolis, October 2008. As a comparison, the United States consumed more than 142 billion gallons of motor gasoline in 2007, and ethanol has roughly two-thirds the energy per gallon of gasoline. EIA, *Annual Energy Outlook 2008*.

27. U.S. Renewable Fuels Association, "U.S. Fuel Ethanol Industry Biorefineries and Production Capacity, November 2008, http://www.ethanolrfa.org./industry/locations/.

U.S. ethanol growth is supported by government subsidies ($0.51 per gallon), targets for gasoline blending, as well as tariff protection.[28] Ethanol is the preferred substitute for fuel oxygenate methyl tertiary butyl ether (MTBE), which is now banned in 19 of the 50 United States, with pending bans in several more states. Federal legislation has set out aggressive goals for renewable fuels, including advanced biofuels.[29] As such, the demand outlook for domestically produced ethanol—or other gasoline and diesel compatible biofuels—remains very favorable in the United States going forward.

Cellulosic ethanol, often referred to as second-generation ethanol, is an advanced process that harnesses much more of the energy potential contained in biomass. There are two main ways to convert cellulosic feedstock into ethanol fuel—biochemical and thermochemical. The first uses hydrolysis followed by fermentation of freed sugars. This is depicted in Figure 6. Many companies are researching production techniques and enzyme development at this time. The second involves gasification through the Fischer-Tropsch process.

Figure 6. Biochemical Conversion Pathway for Cellulosic Ethanol and Main Research Areas[30]

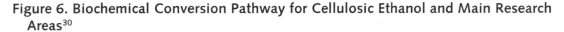

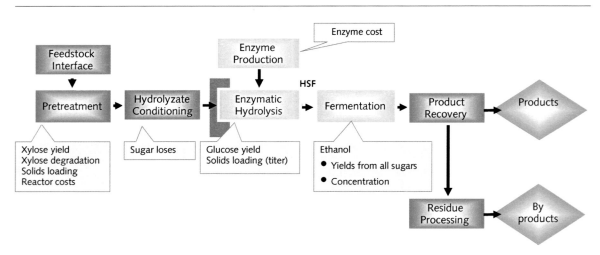

Brazil and the United States are the top two ethanol producers worldwide, producing 89 percent of global supply, with production from China a distant third (Figure 7 illustrates the growth of ethanol production in these countries). In the 1970s, both Brazil and the United States established government programs to support the growth of ethanol as a response to high oil prices. Domestically, the viability of the U.S. ethanol industry has since fluctuated in line with oil prices. While it enjoyed a resurgence of late with sustained high crude oil prices and recognition of the geopolitical and environmental challenges of various alternatives, the impacts of a global economic slowdown and a drastic drop in oil prices have made survival more difficult, in spite of the policy requirements.

---

28. "Zacks Analyst Interview Highlights: Archer Daniels Midland, ExxonMobil, Chevron and Amerada Hess," *Business Wire*, April 3, 2006.

29. See *Energy Independence and Security Act of 2007*, Public Law 110-140, 110th Cong., 2nd sess. (December 19, 2007).

30. Dan E. Arvizu, "Opportunities and Challenges for Alternative Fuels" (presentation, CSIS, Washington, D.C., June 15, 2006), http://www.csis.org/media/csis/events/060615_arvizu.pdf.

As shown in Figure 7, Brazil produced 13.5 million tons of oil equivalents, about 6.9 billion gallons,[31] of ethanol in 2008, utilizing about half its annual sugarcane crop. The United States produced 17.5 million tons of oil equivalents of ethanol in 2008, up 41 percent year-on-year, utilizing about 30 percent of its corn crop.[32] China produced 1 million tons of oil equivalents of ethanol in 2008, down 2 percent over 2007, utilizing a small percentage of its corn and wheat crops. European ethanol production grew by 50 percent in 2008 to 1.3 million tons of oil equivalents to continue to meet demand as more governments have decided to introduce mandatory gasoline blending targets and cut taxes on ethanol.[33]

Figure 7. Global Ethanol Production in the United States, Brazil, and Rest of the World[34]

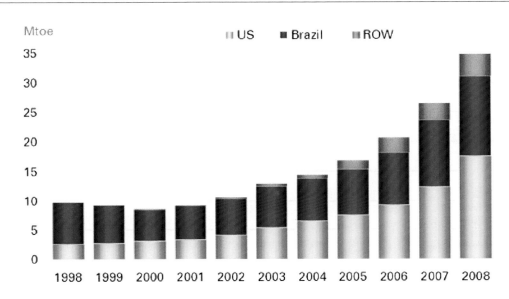

Biofuels are destined to play an important role in the United States even if they never account for a significant portion of total liquids demand. Given strong government support, the demand outlook for domestically produced ethanol—or other gasoline and diesel compatible biofuels—remains very favorable going forward.

Expanding the scale of bioethanol, however, remains an issue. Current estimates indicate that 15 billion to 16 billion gallons per year of corn ethanol is physically feasible without significantly impacting food production.[35] Significant increases in the price of corn feedstock, unanticipated cross-sector impacts on food, international outcry on export prices, and water availability are be-

31. Metric tonne ethanol = 7.94 petroleum barrels = 1,262 liters ethanol energy content (LHV) = 11,500 Btu/lb = 75,700 Btu/gallon = 26.7 GJ/t = 21.1 MJ/liter. HHV for ethanol = 84,000 Btu/gallon = 89 J/gallon = 23.4 MJ/liter ethanol density (average) = 0.79 g/ml (=metric tonnes/m3).

32. Ethanol production used 91.4 million of 304 million metric tons in 2008. National Corn Growers Association, "World of Corn: 2009 Statistics Book," Chesterfield, Mo., 2009, http://ncga.com/files/pdf/WOC2009MetricStatBook.pdf.

33. Ethanol production data and percentage change for Brazil, the United States, China, and Europe are from BP, *Statistical Review of World Energy 2009* (London: BP, 2009), http://www.bp.com/productlanding.do?categoryId=6929&contentId=7044622.

34. BP, *Statistical Review of World Energy 2009*, http://www.bp.com/sectiongenericarticle.do?categoryId=9023791&contentId=7044194.

35. Government Accountability Office, "Biofuels," June 2007, http://www.gao.gov/new.items/d07713.pdf.

ginning to slow corn-based ethanol expansion in the United States. Further, the current economic crisis has forced some producers into bankruptcy. Second generation, cellulosic ethanol technologies, however, offer the promise of increased biofuels production capacity without the food versus fuel trade-off and are receiving considerable funding for initial commercialization, including from the U.S. Department of Energy, venture capitalists, and many state-level incentives. However, the Department of Energy has only just begun to award the loan guarantees that the Energy Policy Act of 2005 authorized them to grant.[36] As of March 20, 2009, DOE announced the first loan guarantee to Solyndra, Inc., for $535 million.[37]

A variety of processes have been shown to be effective at pretreating lignocelluloses, including processes based on exposure to dilute acid, steam, hot water, ammonia, lime, and other agents. Many of these developments are within the Energy Department's biofuels program, which has laid out aggressive cost-reduction targets over the next decade, building on the initial success of driving enzyme costs down. The projected price of production of cellulosic ethanol is currently around $3.00/gallon of gasoline equivalent (GGE) with expectations that it will continue to drop. By 2012, it is expected that this cost will decrease by a factor of two to approximately $1.42/GGE.[38] The most promising aspect of cellulose-to-fuel is that the feedstock base is much broader, including switchgrass, woodchips, straw, and agricultural waste products. Although there are no large-scale commercial facilities producing cellulosic ethanol at this time, large R&D and commercialization efforts, led by the Department of Energy and strongly supported by private corporations, investors, and states, are currently being pursued. The first demonstration and commercial-scale production facilities are now being built, and projections suggest it will be possible to achieve cost reductions approaching $1/GGE and scale of 60 bgal/yr (3.9 mpbd on volume, or about 2.7 mbpd GGE) in the medium term (e.g., 2030). As depicted in Figure 8, the U.S. Department of Energy is targeting substantial decreases in production cost and feedstock cost (per delivered gallon) as well as continued cost reduction of the enzymes.

Internationally, Brazil is the lowest cost and, according to many accounts, most environmentally sound, worldwide producer of ethanol, primarily sourced from domestically grown sugar cane.[39] Petrobras (Brazil's national oil company) is looking to invest some $1.5 billion over the next five years on pipelines and ships to help boost export marketing.

36. Government Accountability Office, "Department of Energy: New Loan Guarantee Program Should Complete Activities Necessary for Effective and Accountable Program Management," July 2008, http://www. gao.gov/new.items/d08750.pdf.

37. U.S. Department of Energy, Office of Public Affairs, "Obama Administration Office $535 Million Loan Guarantee to Solyndra, Inc.," March 20, 2009, http://www.lgprogram.energy.gov/press/032009.pdf.

38. Dan E. Arvizu, "Opportunities and Challenges for Alternative Fuels" (presentation, CSIS, Washington, D.C., June 15, 2006), http://www.csis.org/media/csis/events/060615_arvizu.pdf.

39. There is considerable debate on the methods and details for evaluating the full environmental aspects of biofuels. Full life-cycle analysis that includes land-use impacts, as well as production and processing, depends strongly on the details of farming practices, processing, and distribution, as well as the overall production amount of the fuel. For example, see presentations at the Global Bioenergy Partnership, 2nd Task Force on GHG Methodologies, March 6–7, 2008, Washington, D.C., http://www.globalenergy.org/ events1/gbep-events-2008/task-force-on-ghg-2008/en/.

Figure 8. Cost Production Targets for Cellulosic Ethanol[40]

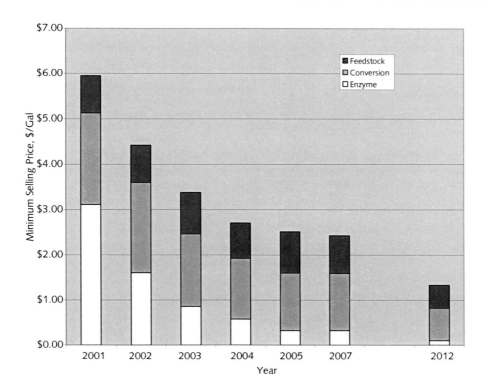

Biobutanol

Butanol, like ethanol, is produced through the fermentation of biomass, whereby bacteria to yield the alcohol break down feedstock. Corn, wheat, sugar beet, sugar cane, cassava, and sorghum are all usable feedstock. As an alternative transportation fuel, butanol is also compatible with current engine technology (in certain blends) and useful as either a supplement to or substitute for ethanol.[41]

While it is not yet commercially available as a fuel for transportation, an established U.S. market already exists for butanol used as an industrial solvent. Biobutanol advocates estimate that demand over the past few years has been around 350 million gallons at a wholesale cost of about $3.35 per gallon.[42] Butanol is insoluble in water and will resist separation. This characteristic

40. EIA, "Impacts of Energy Research and Development (S.1766 Sections 1211-1245, and Corresponding Sections of H.R.4) with Analysis of Price-Anderson Act and Hydroelectric Relicensing," http://www.eia.doe.gov/oiaf/servicerpt/erd/renewable.html.

41. The energy content of butanol is 105,000 BTU/Gal or 29.2 MJ/L, slightly less than that of gasoline. This means an equal volume substitution of butanol for gasoline would yield 9 percent less energy, which corresponds proportionally to a shorter vehicle range. Because the characteristics of butanol fuel are more similar to gasoline compared with ethanol, it can be blended without engine modification at a slightly higher volumetric ratio, 11.5 percent, in the United States than ethanol under current regulation. One disadvantage of butanol compared with other fuels is its lower octane rating range, 78 to 96, which results in greater engine knocking during fuel burn and reduced efficiency.

42. Environmental Energy, Inc., letter to Dennis Weaver and Others, April 30, 2004, http://nmhemp.org/A_Butanol_Economy.pdf.

enables distribution of the fuel through current infrastructure, including pipelines, with no additional modifications.

Several major companies, notably BP and DuPont, are actively pursuing butanol research. These companies announced a joint venture in 2006 to retrofit a UK ethanol plant to produce 9 million gallons per year of butanol from sugar beet feedstock and demonstrate its viability as a fuel additive in the British market.[43] While the companies' 2008 projections estimated that their advanced biobutanol could compete economically with ethanol by 2010, more recent statements indicate that the substance will not be produced in large quantities until 2013.[44]

Biodiesel

Oilseeds, like rapeseed and soy, can be extracted and converted to esters and are well suited to replace diesel at limited scale. Rapeseed-based biodiesel production is fully commercial in Europe.[45] Significant quantities of rapeseed methyl ester (RME) are produced in the European Union (concentrated in Germany, France, and to a lesser extent in Austria and Italy). Subsidies for RME production in Europe generally consist of a combination of farm subsidies (e.g., for producing nonfood crops) and tax exemption of the fuel itself. Inclusive subsidies for RME are 300 percent to 400 percent greater than those for conventional diesel or gasoline production.[46] Key drivers for the implementation of RME schemes are rural employment and the flexible nature of the crop because it can easily replace conventional food crops when desired.

The rapidly changing character of biodiesel production capability is illustrated by recent trends in the United States. In 1995, U.S. biodiesel production was 1.9 million liters (500,000 gallons). By 2004, it had jumped to 95 million liters (25 million gallons), and by 2005, it had again tripled, to more than 280 million liters (75 million gallons).[47] Production in 2006 and 2007 was 250 and 450 million gallons, respectively.[48] Most U.S. production is soy based, and it consumes about 10 percent of the soy crop.[49]

Palm oil–based biodiesel is popular in palm-growing nations (notably Malaysia and the Philippines), but the recent increase in fuel production that led to unanticipated environmental impacts has forced governments to reconsider this source and to establish environmental sustain-

43. BP, "BP, ABF, and DuPont Unveil $400 Million Investment in UK Biofuels," June 26, 2009, http://www.bp.com/genericarticle.do?categoryId=2012968&contentId=7034350.

44. Clifford Krauss, "Big Oil Warms to Ethanol and Biofuel Companies," *New York Times*, May 26, 2009, http://www.nytimes.com/2009/05/27/business/energy-environment/27biofuels.html?_r=2&ref=energy-environment.

45. André Faaij, "Assessment of the Energy Production Industry: Modern Options for Producing Secondary Energy Carriers from Biomass," in *Renewables-Based Technology: Sustainability Assessment*, ed. Jo Dewulf and Herman Van Langenhove (New York: Wiley, 2006).

46. Faaij, "Modern Biomass Conversion Technologies."

47. Market Research Centre, "Agri-Food Trade Service: The Biofuels Market in the U.S. Upper Midwest," International Trade Canada, June 2006, http://atn-riae.agr.ca/us/4181_e.htm.

48. "Coping with the added pressure of the RFS," *Biofuels International* 2, no. 1 (March 25, 2008), http://www.biofuels-news.com/articles/v7_a3.html.

49. Martin Tobias, "Biodiesel to Take up 10% of US Soy Crop this Year," Cleantech Blog, August 2, 2007, http://news.cnet.com/8301-13511_3-9754205-22.html?hhTest=1.

ability criteria.[50] Long-term prospects for material quantities of biodiesel from tropical countries do not appear practical.

Some discussion and early research results of jatropha as a bioresource indicate appropriateness for rural, arid climates. Initial project development along with pilot-scale feasibility analysis has begun in India, Indonesia, and various African countries. In late 2008, Air New Zealand expressed its support for the biofuel by utilizing a jatropha–jet fuel blend on a test flight.[51] In India, BP has made a $9-million R&D investment while some biofuel companies have invested in large-scale cultivation, more than half a million acres.[52] Despite these signs of investor interest, scale issues related to biodiversity and the economics of collection, processing, and transportation to demand markets have yet to be fully addressed.

Four principle technology pathways are in various stages of maturity: esterification, pyrolysis, hydrogenation, and gasification. First generation esterification is commercial today, with incremental advancements being made via process changes such as reverse flow, acid/base pretreatments, and the use of other catalysts. Pyrolysis continues at pre-commercial scales and is faced with multiple technical challenges. Hydrogenation is emerging as a pathway to formulate blends of bio-derived fuels and petroleum fuels. Gasification of pure bioresources for diesel compatibles is not yet economically commercial, but co-fired fossil/bio plants appear to offer some upsides in terms of costs, product quality, and flexibility.

Depending on the operating conditions (temperature, heating rate, particle size, and solid residence time, etc.), pyrolysis can be divided into three subclasses: conventional, fast, or flash. To maximize bio-oil production, fast/flash pyrolysis is used, heating the biomass at about 500°C for less than 10 seconds. Biomass pyrolysis faces numerous technical issues to be economic at further scale. Pyrolysis oils contain suspended char and alkali metals that require downstream processing. Additional challenges include acidity and water content. While cost effective treatments exist for use in combustion boilers, stationary diesel engines, or industrial combustion turbines, few options are in use today for upgrading these oils for use as a commodity transportation fuel.[53]

It should be noted that biomass could also be converted into liquid oil by direct hydrothermal liquefaction, which is a form of intermediate pyrolysis that occurs in the presence of water. Many other variants of pyrolysis are under development. Overall, to date, pyrolysis has not been proven either technically or economically viable for producing transport, quality-grade fuel.

A number of recent commercial announcements indicate growing interest in processing biomass inputs into intermediate products that are compatible with existing production infrastructure and that blend well with petroleum diesel. The fundamental approach is to process biofeedstocks into non-oxygen-containing diesel distillates. New processing approaches have been developed to produce a bio-based "renewable" or "green" diesel that has no oxygenate produc-

50. See, for example, proposals from Switzerland and Germany at the Global Bioenergy Partnership, 2nd Task Force on GHG Methodologies, March 6–7, 2008, Washington, D.C.

51. James Kanter, comment on "Air New Zealand Flies on Engine With Jatropha Biofule Blend," Green Inc. blog, *New York Times*, December 30, 2008, http://greeninc.blogs.nytimes.com/2008/12/30/air-new-zealand-flies-on-engine-with-jatropha-biofuel-blend/?scp=1&sq=jatropha&st=cse.

52. Jon R. Luoma, "Hailed as a Miracle Biofuel, Jatropha Falls Short of Hype," *Guardian*, May 5, 2009, http://www.guardian.co.uk/environment/2009/may/05/jatropha-biofuels-food-crops.

53. Suzanne Hunt, "'Biofuels for Transportation'" Report Findings—Online Discussion," Worldwatch Institute, June 14, 2006, http://www.worldwatch.org/node/4077; and S. Czernik and A.V. Bridgwater, "Overview of Applications of Biomass Fast Pyrolysis Oil," *Energy & Fuels* 18, no. 2 (March 2004): 590–598.

tion, zero sulfur, nitrogen, or aromatics, and excellent compatibility with oil-refined diesel, with little retrofit hurdles. The first to announce the "integrated refinery oil hydrogenation process (IROHP)" was Neste Oil of Finland, which opened a 10,000 bpd (170,000 Mt/yr) unit in Porvoo in summer 2007 using its NExBTL technology. Neste has plans for two larger (800,000 Mt/yr) facilities; one in Singapore due to be completed by the end of 2010 and another in Rotterdam in 2011.[54] In the last few years, several others, including Petrobras (H-Bio), BP, Conoco Phillips, Nippon Oil, and others have announced similar developments. Petrobras adapted four refineries to produce H-Bio but did not begin mass production as planned in January 2008 due to the high price of soy oil, which at the time was about $180/barrel.[55]

It appears that an 80/20 blend of oil from petroleum/vegetable oil offers processing advantages in existing refineries. This not only results in a higher quality "renewable diesel" product, but also significantly improves process and product economics (albeit the product is not 100 percent "bio based"). Verification of the biomass content is a consideration for meeting alternative fuel criteria.

Gasification of biomass into synthesis gas (a combination of carbon monoxide and hydrogen) occurs at high temperatures under strictly controlled conditions that are specific to the feedstock, technology, and pretreatment. Given the complex nature of many biofeedstocks, controlling the process and "cleaning" the gas for use in further processing steps (e.g., FT fuel synthesis) remains a major challenge. Meanwhile, gasification using fossil fuel feedstock (primarily coal and petroleum) rather than biomass is already commercially established technology for the production of both electricity and liquid fuels. Approximately 120 plants operating worldwide have generated a variety of outputs, including chemicals (produced in 37 percent of plants), Fischer-Tropsch liquids (36 percent), power (19 percent), and gaseous fuels (8 percent). More than 35 new plants are planned to come online by 2010, but none is currently focused solely on biomass feedstocks.[56]

It is beyond the scope of this report to address all opportunities to combine fossil fuels with coproducts. In principle, gasification-based conversion platforms offer flexible fuel production from biomass and coal as well as natural gas. CTL fuel production can also become more commercially attractive if carried out in combination with electricity generation, a coproduct of CTL fuel processing.

Theoretical studies have shown that about 50 percent of the carbon in gasified coal can be captured when producing Fischer-Tropsch liquids via gasification. When more biomass is utilized, negative emissions could be obtained, depending on the source and carbon life-cycle properties of the biomass feedstock.[57] The concept, not yet proven, requires CO_2 capture, in combination with "net negative" biomass production, as shown in Figure 9.

In combination with biomass co-gasification and with the addition of carbon capture and stor-

54. Neste Oil, "NExBTL Diesel," June 13, 2008, http://www.nesteoil.com/default.asp?path=1,41,535,547, 3716,3884.

55. Green Car Congress, "Petrobras H-Bio Production on Hold," Reuters, January 19, 2008, http://www. greencarcongress.com/2008/01/petrobras-h-bio.html.

56. Worldwatch Institute, *Biofuels for Transport: Global Potential and Implications for Sustainable Agriculture and Energy in the 21st Century* (London: Earthscan Publications, 2007), 71.

57. Ibid.

age, co-fueled polygeneration plants have been conceptualized and evaluated, leading to possible reduction of carbon emissions while simultaneously introducing an additional business opportunity in CTL/BTL fuel production.

Figure 9. Process Flow of Coal and Biomass Resource Supplied to Polygeneration Plant with Carbon Capture and Storage[58]

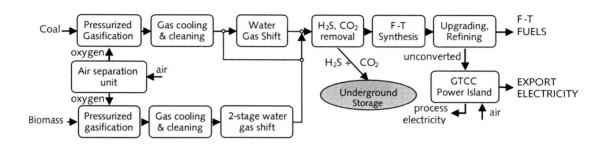

Opportunities and Challenges

There are many challenges to increasing alternative fuels production and use in the United States. Challenges facing supply growth include competition for food supply, land-use impacts, carbon impacts, water, feedstock production, technology maturity, and international trade and security.

Infrastructure and Distribution

The global oil market is the world's largest supply chain, and the scale of oil consumption is unprecedented: 3 billion gallons of liquid fuel are used daily. This works out to half a gallon of fuel per human per day. The current system, which took over a century to develop, includes exploration, extraction, refining, production, distribution, and marketing and at each point is under pressure to expand to meet the anticipated growth in global demand over the decades ahead. There exist many opportunities for alternative fuels to alleviate some of the pressures on the system, but the notion of large-scale replacement anytime soon is more political rhetoric than fact—even under optimistic conditions.

Massive amounts of capital would be required to introduce new technologies and feedstock into the supply chain at significant scale. New alternative and supplemental fuels require new supporting infrastructure not limited to production facilities and a distribution network. Recent estimates project that $7 trillion is needed over the next few decades for investment in alternative energy to meet climate change mitigation targets.[59] Globally, this figure rises to $17 trillion to stabilize emissions or $45 trillion (1.2 percent of world GDP) to halve emissions by 2050.[60]

---

58. Robert Williams, "Alternative Fuels Seminar: Carbon to Liquids" (presentation, CSIS, Washington, D.C., December 12, 2006), http://www.csis.org/media/csis/events/061212_williams.pdf.

59. Cambridge Energy Research Associates (CERA), "Global Climate Change Response Can Spur $7 Trillion in Clean Energy Investment by 2030: CERA Analysis," press release, February 5, 2008, http://www.cera.com/aspx/cda/public1/news/pressReleases/pressReleaseDetails.aspx?CID=9239.

60. International Energy Agency (IEA), *IEA Energy Technology Perspectives 2008* (Paris, France: IEA, 2008).

Table 2: Infrastructure Costs and Estimated Benefits for Various Alternative Fuels[61]

Item	Ethanol	Methanol	CNG	Hydro-gen	Fischer-Tropsch	Comments
Station cost for conversion ($1,000)	170	182	926	1,423	N/A	Values are cost per station
2030 high case annual capital cost ($ billion)	10.44	9.16	233	59.56	1.22	
2030 fuel cost ($/gallon)	1.95	1.60	1.70	4.84	1.02	Consumption 50,000 gal./mo. gasoline equivalent
2030 Reduction in GHG emissions (MMT CO_2)	678	374	427	496	343	

Alternative fuels have a different risk profile than that of traditional petroleum businesses, and the risk profile for biofuels differs from that for unconventional fuels. Biofuel supply will vary depending on weather and crop availability, and political forces may limit its growth depending on reaction to cross-sector economic impacts (including geopolitical issues related to trade and cross-border economics). The risk profile for unconventional fuels is more similar to that for conventional oil, but the high cost of production could limit its viability at times of lower oil or higher natural gas prices, and its often-elevated environmental impact may make it vulnerable to shifting political winds.

Even without consideration of new alternative transportation fuels, the capacity of all freight-transportation options is currently becoming constrained. Additional freight of liquid biofuels will only further strain the system and of course is less than optimal on a per unit basis. As such, significant strategic issues related to the dispersed nature of alternative fuel feedstock, processing facilities, and demand centers remain to be addressed as the scale of alternative fuel production and use grows.

With synthetics and unconventional resources, there is a strong case for manufacturing at or very near the resource base. This is because while some of the new fuels, such as synthetic oil shale crude from Alberta, are rather easy to plug into the system, others, like biofuels, may require the creation of entirely new production and distribution chains. For example, coal traditionally moves by rail to its point of usage. If the production of coal doubles for CTL processing, there will be an increased demand on an already strained railroad network to transport the resource from the mine mouth to the processing facility. If the CTL plants are built at mine mouths, there will be a

---

61. Committee on Alternative Transportation Fuels, http://onlinepubs.trb.org/onlinepubs/millennium/00005.pdf.

need for more pipelines. The distribution map shown in Figure 10 and the map of existing and planned biofuel plants in Figure 11 provide a visual contrast of the existing infrastructure. Meanwhile, the bulk of domestic production of biofuels takes place in the Midwest, whereas the largest demand centers are located along the East and West Coasts.

Figure 10. Major Refined Product Pipelines[62]

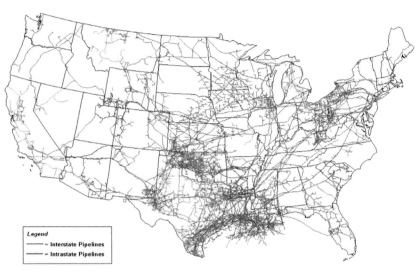

Legend
——— = Interstate Pipelines
——— = Intrastate Pipelines

Source: Energy Information Administration, Office of Oil & Gas, Natural Gas Division, Gas Transportation Information System

The economics of biofuels indicate an optimum size of perhaps 100 million gallons per year (6,500 b/d), driven primarily by transportation costs of low-density feedstock versus, for example, a 1 million to 2 million b/d or more refinery. It is most likely that a geographic dispersion of production plants will emerge, thus forcing increased focus on transportation of the produced fuel. Today, biofuels transport is mainly accomplished by truck and rail. Only one dedicated biofuels pipeline exists today, and there are considerable challenges to adapting existing infrastructure for large-scale biofuels transport.

---

62. EIA, Office of Oil & Gas, Natural Gas Division, Gas Transportation Information System, "U.S. Natural Gas Pipeline Network, 2009," http://www.eia.doe.gov/pub/oil_gas/natural_gas/analysis_publications/ngpipeline/ngpipelines_map.html.

Figure 11. Location of Existing U.S. Bioethanol Refineries[63]

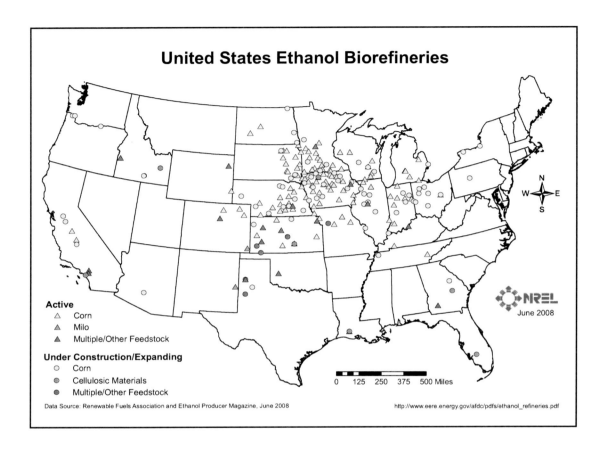

Water, Land, and Carbon

Environmental concerns are a major driver behind the current focus on increasing production of biofuels both in the United States and globally. As alternatives gain more attention, they also receive more scrutiny. Comprehensive studies on the environmental benefits vary with regard to assumptions and methodologies. While carbon emissions dominate the environmentally oriented discussion, it is important to remember that the environment extends beyond greenhouse gas emissions to water, energy requirements, and land usage. The full life-cycle impacts of new alternatives must be considered when weighing the benefits of increased production and use. Moreover, the multidimensional nature of energy forces policymakers to find solutions that balance environmental, economic, and security policy goals. There may be no single, ideal solution, but rather a portfolio of solutions combined with a comprehensive reassessment of policy goals.

While technology and new processes for accessing unconventional fossil fuels such as oil shale and oil sands continue to be improved, traditional extraction techniques require the resources to be mined—much like coal—and consequently carry negative impacts for the surrounding environment. All unconventional fuel sources require water for their production and emit greenhouse gases. The water-use impacts and CO_2 emissions must be compared with those for conventional

---

63. U.S. Department of Energy, Alternative Fuels and Advanced Vehicles Data Center, http://www.afdc.energy.gov/afdc/pdfs/ethanol_refineries.pdf.

oil. Refining oil requires between 1 and 2.5 gallons of water per gallon of refined product, and CO_2 emissions are 8,800 to 10,000 grams per gallon, for gasoline and diesel fuel respectively, according to the Environmental Protection Agency.[64] Lifecycle water use for the four options discussed is generally at least double that for conventional oil, although natural gas-to-liquids may use less.[65] Without carbon capture and storage (CCS), CO_2 emissions for the four options, except for natural gas-to-liquids, are between 30 and 100 percent higher than that for conventional oil. They are relatively comparable with CCS, but CCS generally increases water demands.[66] From a broad perspective, the increase in CO_2 is likely more troubling than the increase in water use because oil accounts for about 40 percent of U.S. CO_2 emissions, but only about 1 percent of domestic water consumption is used for oil refining. However, plants for some unconventional fuels such as coal-to-liquids would need to be extremely large to be economical (i.e., about 1 billion gallons/year per facility compared with 50 million to 100 million gallons/year for a typical ethanol plant), and the impact on local water consumption could be very significant.

Additionally, water use is increasingly becoming a critical factor as new operations are considered, yet it is often overlooked when discussing alternative fuels. While this series did not focus on global water resources, there are myriad studies of water use and supply in the conversion process, as well as with regard to specific geographic projects in areas where water is in short supply.[67] All alternative fuels require water in the conversion process, usually between two and three times as much as the water needed to refine a gallon of gasoline. Whereas water consumption is estimated at between 1 and 2.5 gallons of water per gallon of gasoline (gal/gal), coal-to-liquids, hydrogen, and ethanol all need 4.5 to 7 gallons of water per gallon of gasoline energy-equivalent product. Shale (not in situ) needs 2 to 5 gal/gal for oil extraction alone in addition to the water needed for refining. (Note: processing biodiesel requires less than 1 gal/gal.) Such an increase in water consumption will exacerbate water shortages in some locations, but considering that oil refining currently only consumes about 1 percent of the nation's water, the effect will be geographically concentrated.

In addition, there is a large quantity of water that is polluted during the mining of fossil fuels and from agricultural production of biofuels crops. The rapid expansion of corn grown for ethanol, which requires large quantities of nitrogen and phosphorous, increases runoff pollution from fertilization. Water is one of the critical issues of this century, and more comprehensive evaluations of water resources usage and impacts for alternative fuels are needed.

Biomass production also requires land. The estimated productivity for perennial energy crops like willow, eucalyptus, or switchgrass is reported to be between 8 and 12 tons of dry matter per hectare per year. A comprehensive review of 17 studies of biomass availability showed the large

---

64. U.S. Environmental Protection Agency, "Emission Facts: Average Carbon Dioxide Emissions Resulting from Gasoline and Diesel Fuel," February 2005, http://www.epa.gov/OMS/climate/420f05001.htm.

65. Secondary recovery of oil does use a significant amount of water, but oil production results in produced water from wells. While it varies by well, in general about three barrels of produced water comes to the surface with each barrel of oil. Much of this water is simply injected back, potentially leading to no additional consumption during secondary recovery.

66. In some cases, renewable electricity could be used that does not require as much water.

67. For further reading, please see, U.S. Department of Energy, "Energy Demands on Water Resources," December 2006, http://www.netl.doe.gov/technologies/coalpower/ewr/pubs/DOE%20energy-water%20 nexus%20Report%20to%20Congress%201206.pdf.

range of projected total bioenergy supply.[68] Figure 12 summarizes the main findings. The authors found that the possible contribution of biomass in the future global energy supply ranged from below 100EJ yr−1 to above 400 EJ yr−1 in 2050. The results were shown to be highly dependent on four factors: land availability, yield levels in energy crop production, forest wood, and agriculture residue estimates, all of which are very uncertain and subject to widely different opinions.

Figure 12. Results of a Review of 17 Studies Projecting Biomass Potentials up to the Year 2100, Expressed in EJ/yr[69]

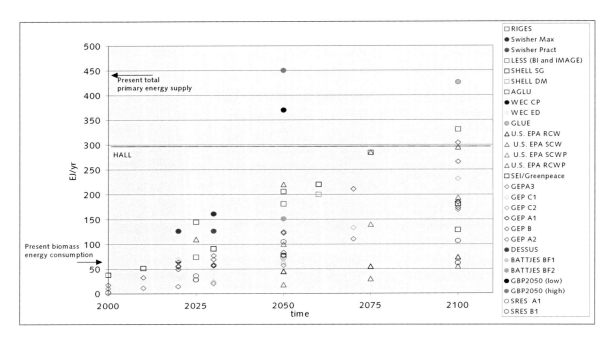

There were a few consistent results, however, including the notion that both the technical and economic potential of biomass resources for energy and material use can be very large (up to more than two times the current global energy demand). To reach these potentials, it was argued that the development of competitive energy cropping systems would require the "rationalization" of agriculture, especially in developing countries. Perennial crops (such as eucalyptus, poplar, grasses such as miscanthus, and sugar cane) appear to provide the most favorable economics and environmental characteristics for biomass production, depending on land-use change and farming practices.

For biofuels with dedicated crops, new farming practices and breakthroughs in cellulosic processing will relieve some of the pressure raised by recent studies on land/water/carbon impacts. Under current assumptions related to lifecycle analysis, biofuels generally appear to have lower CO_2 emissions. However, biodiversity loss is also a critical issue in both the United States and Europe, where policymakers are rethinking aggressive biodiesel policies due to the previously unanticipated devastation of tropical forests by farmers planting palm trees. As a consequence, an overdue and serious discussion of "sustainable certification practices" is now underway.

---

68. Göran Berndes et al., "The Contribution of Biomass in the Future Global Energy Supply: A Review of 17 Studies," *Biomass and Energy* 25 (2003): 1–28.
69. Ibid.

Significant advances are still required to exploit this bioenergy potential, including the ability to improve crop yields globally, especially in developing countries.[70] Regional differences also affect the extent and speed of transitions to bioenergy. Numerous scenarios, including drought, natural disasters, restricted trade, and uncertain government policies could lead to bioenergy volumes that are extremely low. The total energy supply potential for energy crops therefore largely depends on land availability and must consider global growth in demand for food, environmental protection, sustainable management of soils and water, and a variety of other sustainability criteria. These and many other assumptions that drive long-term potential estimates make it impossible to present the future biomass potential in one simple figure.

There has been much intense debate focused on the carbon balance of various alternative transportation fuels. Conflicting studies have been published, each exhibiting a variety of differing assumptions. The main question to be addressed is whether or not alternatives offer a net gain over petroleum in terms of life-cycle carbon emissions.

Each feedstock and technology pathway must be evaluated comprehensively at scale. Options of the various land use, farming practices, conversion processes, transportation, and distribution all must be accounted for. For ethanol, there are several contradictory studies regarding the impact that ethanol will have in reducing greenhouse gas (GHG) emissions compared to petroleum. There appears to be a consensus (with a few outliers) that sugar cane and cellulosic ethanol have considerable GHG reduction potential over gasoline and unconventional fossil fuels, but the case for corn ethanol is less so. The numeric values placed on making the switch to ethanol are widespread and difficult to decipher.[71] Part of the reason for the conflicting data is due to the varying assumptions made during calculations (e.g., management practices, conversion and valuation of coproducts). Because each study makes different assumptions, the conclusions are far from uniform.[72]

One such study looked at six analyses of fuel ethanol from corn.[73] The study found that switching from gasoline to corn ethanol generated a range of anywhere from a 20 percent increase to a 32 percent decrease in emissions. However, when comparing cases from the study, corn-based ethanol versus cellulosic (assumes that production from switchgrass becomes economical), the findings show that cellulosic ethanol can drastically reduce emissions. The study reported that ethanol from corn produces 77 g CO_2e per MJ while ethanol from cellulosic material produces significantly less, only 11 g CO_2e per MJ compared to gasoline, which produces 91 g CO_2e per MJ.

Other results are generally consistent with those above, with specifics that depend on a broad set of assumptions, ranging from land origin and use, farming practices, water use and rainfall (geography), through processing technology and distribution.[74] Interestingly, most analyses to date, particularly for fermentation pathways, ignore the possibility that the fermentation step is a pro-

70. Faaij, "Modern Biomass Conversion Technologies."

71. See, for example, the range of analyses presented at the Global Bioenergy Partnership, 2nd Task Force on GHG Methodologies, March 6–7, 2008, Washington, D.C.

72. See, for example, Timothy Searchinger et al., "Use of U.S. Croplands for Biofuels Increases Greenhouse Gases through Emissions from Land-Use Change"; and rebuttal, M. Wang and Z. Haq, letter to the editor, *Science*, March 14, 2008, http://www.transportation.anl.gov/pdfs/letter_to_science_anl-doe_03_14_08.pdf.

73. Alexander E. Farrell et al., "Ethanol Can Contribute to Environmental and Energy Security," *Science* 311, no. 5760 (January 27, 2006): 506–508.

74. Results of a series of analyses are available at Global Bioenergy Partnership, 2nd Task Force on GHG Methodologies, March 6–7, 2008, Washington, D.C.

cess where pure CO_2 is produced and captured and could be deployed relatively easily. Combining the use of sustainably grown biomass with (partial) CO_2 capture would allow for overall negative CO_2 emissions per unit of energy produced on a life-cycle basis.[75]

Technological developments throughout the supply chain, from production to harvesting to densification to transportation of feedstocks to conversion and transportation and distribution of final products will be needed to improve the competitiveness and efficiency of bio-based replacements for petroleum.[76] While market growth and technological advances can help reduce costs, it is clear that biofuels should be viewed as a welcome supplement instead of a large-scale replacement to current fuels in the energy supply picture of the coming decades.

In summary, estimates of the potential total market share of biofuels very widely, but perhaps more importantly, nearly all estimates indicate that the maximum potential of biofuels to the global liquid-fuel economy in the next 20 to 30 years will be limited to less than 50 percent and more likely 10 to 20 percent. Regional specifics present an even broader range of challenges, given available land, food production requirements, water availability, and challenges of estimating non-crop resources such as municipal solid waste, forest and agriculture wastes, and waste oils. Sustainability requirements may force slower and more careful growth, which in turn may reduce the overall potential contribution of biofuels.

Across the fuel supply landscape, and more broadly across the energy for transportation landscape, multiple options are emerging as technically feasible and politically, economically, and environmentally attractive. Alternative fossil fuel supplies are increasingly attractive, especially if world oil prices exceed $100/barrel; albeit with increasing concerns over longer-term environmental issues related to more challenging extraction and conversion needs. Bio-based liquid fuels, while dominated by ethanol and biodiesel today, including utilization of food crops (e.g., corn in the United States), offer a wide variety of technical pathways, many of which are not yet proven at commercial scale. Moreover, any "alternative" option—including the use of electricity for electric vehicles or plug-in electric vehicles—faces the issues of scale across the supply chain, from resource availability to conversion to distribution. Multiple creative solutions that comprehensively address the environment, economics, and geopolitical issues will likely have to be pursued in parallel to address the enormous scale of the world transportation sector today and over the next century. Complementary auto technologies and policy environments will be required to complete the portfolio.

75. See, for example, Christian Azar et al., "Carbon Capture and Storage from Fossil Fuels and Biomass: Costs and Potential Role in Stabilizing the Atmosphere," *Climatic Change* 74, nos. 1–3 (January 2006): 47–79.

76. Faaij, "Assessment of the Energy Production Industry: Modern Options for Producing Secondary Energy Carriers from Biomass."

VEHICLE TECHNOLOGIES

The internal combustion engine (ICE) has dominated the automobile market for more than a century. When framing new transportation policy, is important to appreciate the scale of the current system—there are nearly 250 million automobiles on the road in the United States today,[1] and annually these vehicles travel 3 trillion miles on the nation's 46,000 miles of public highways.[2] The ICE will continue to be the main technology utilized until electric cars or other replacements are available at scale to the average consumer.

The level of mobility enjoyed by the U.S. driver, although convenient, comes at a cost. Transportation consumption of liquid fuels accounts for roughly 5 percent of personal disposable income as well as approximately one-third of U.S. carbon dioxide (CO_2) emissions, and 60 percent of transportation emissions are from passenger vehicles.[3] The growing awareness about the impact of carbon emissions on climate change has fueled research on ways to make vehicles more energy efficient and has influenced consumer behavior. Nevertheless, vehicle purchase selections are principally driven by economic considerations and lifestyle choices. What really focused the attention of the U.S. driver was the increase in fuel prices through the summer of 2008. Gasoline prices hovering near $4.00 a gallon made Americans acutely aware of their vehicle fuel economy (or lack thereof). A few years of sustained high prices have slowly begun to influence driving habits, transportation mode choices, and purchase behavior; it is yet to be seen whether there will be a reversion of this trend in the face of steeply declining oil prices.

Efficiency

U.S. requirements for vehicle fuel economy are among the lowest in the developed world.[4] In the years following the introduction of corporate average fuel economy (CAFE) standards—a policy

---

1. The federal government counts 100 million currently registered automobiles (personal vehicles) not including publicly owned vehicles (fleets), trucks, trailers, or other vehicle types. Federal Highway Administration, *Highway Statistics 2006* (Washington, D.C.: U.S. Department of Transportation, December 2007), table MV-1, http://www.fhwa.dot.gov/policy/ohim/hs06/htm/mv1.htm.

2. National Surface Transportation Policy and Revenue Study Commission, *Transportation for Tomorrow: Report of the National Surface Transportation Policy and Revenue Study Commission* (Washington, D.C., December 2007), http://www.transportationfortomorrow.org/final_report/.

3. Brent D. Yacobucci, *Regulation of Vehicle Greenhouse Gas Emissions: State and Federal Standards* (Washington, D.C.: Congressional Research Service, January 11, 2008), http://assets.opencrs.com/rpts/RS22788_20080111.pdf. Disposable income calculation/data taken available at:

4. Fuel economy standards in the United States have historically lagged behind other countries. Feng An and Amanda Sauer, *Comparison of Passenger Vehicle Fuel Economy and GHG Emission Standards Around the World* (Washington, D.C.: Pew Center on Global Climate Change, December 2004), http://www.pewclimate.org/global-warming-in-depth/all_reports/fuel_economy. In the United States, the Energy Independence and Security Act of 2007 raised CAFE (corporate average fuel economy)

response to the oil supply shocks of the 1970s—vehicle manufactures responded by producing smaller, more efficient cars. Over the next decade, oil demand declined, prices dropped, and further increases in fuel economy standards were not as stringent. Average U.S. vehicle efficiency increased throughout the 1980s and then began to slowly decrease over the next decade.

Figure 13. Distribution of New Light-Duty Sales, by Year, Type, and EPA Weight[5]

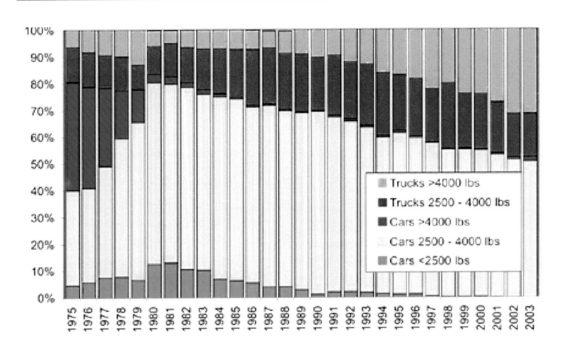

While initial gains in efficiency went to increasing net vehicle fuel economy, further advancements allowed for the production of larger and more powerful vehicles as the U.S. federal government lagged behind in increasing fuel economy standards.[6] The average U.S. vehicle today can accelerate faster than the average model produced 30 years ago while weighing more and housing a

---

standards for passenger vehicles and light trucks from 27.5 mpg to 35 mpg by 2020. *The Energy Independence and Security Act of 2007*, HR 6, 110th Cong., 1st sess. (January 4, 2007), http://energy.senate.gov/public/index.cfm?FuseAction=IssueItems.Detail&IssueItem_ID=f10ca3dd-fabd-4900-aa9d-c19de47df2da&Month=12&Year=2007. The European Union regulates CO_2 emissions from vehicles rather than fuel economy directly. Current vehicles in Europe emit roughly 160 g/km. A new resolution adopted in late 2008 strengthens EU vehicle performance standards to an average of 120g of CO_2/km by 2012, and a long-term target for 2020 for the new car fleet of average emissions of 95 g CO_2/km. European Parliament, "European Parliament seals climate change package," December 18, 2008, http://www.europarl.europa.eu/news/expert/briefing_page/44591-350-12-51-20081216BRI44590-15-12-2008-2008/default_p001c005_en.htm.

5. Tom Wenzel and Marc Ross, *Increasing the Fuel Economy and Safety of New Light-Duty Vehicles*, white paper prepared for the William and Flora Hewlett Foundation's Workshop on Simultaneously Improving Vehicle Safety and Fuel Economy through Improvements in Vehicle Design and Materials, September 18, 2006, p. 1, http://eetd.lbl.gov/ea/teepa/pdf/LBNL-60449.pdf.

6. Ibid.

variety of new technologies and electronics designed to improve safety, convenience, and comfort, in addition to being somewhat more fuel efficient.[7] Heavier vehicles, in particular sport-utility vehicles (SUVs), which were classified as trucks, proliferated. It was not until the run-up in oil prices starting in 2005 that consumer preference for vehicles with higher miles per gallon (mpg) ratings, and vehicle fuel economy once again began to climb.

Stronger fuel economy regulations would affect petroleum demand and carbon emissions in the transportation sector. Figure 14 depicts growth in annual gasoline consumption of light-duty vehicles and various policies that could temper or reverse the trend for increased fuel demand. The top curve, reflecting a reference projection, assuming no change in policy or increase in fuel economy, shows marked growth in the size of the fleet (number of vehicles on the road) and vehicle kilometers traveled (VKT) and potential increase in consumption to 813 billion liters per year. The yellow wedge represents the decrease in energy demand possible from efficiency improvements in mainstream technologies (e.g., engines, transmission, materials).

Figure 14. U.S. Light-Duty Fleet Gasoline Consumption Projections: No Change versus Improved and Advanced Technologies[8]

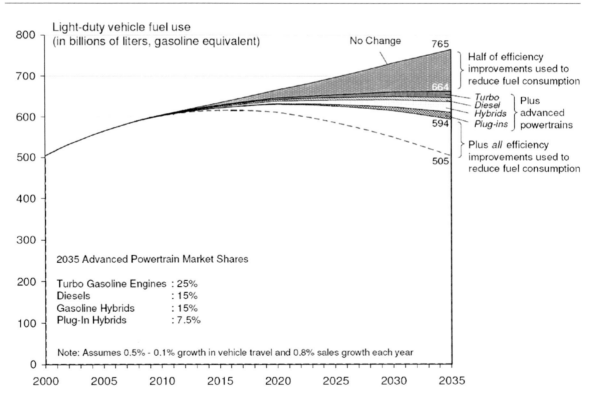

7. U.S. Environmental Protection Agency (EPA), "Light-Duty Automotive Technology and Fuel Economy Trends: 1975 through 2008," September 2008, http://www.epa.gov/OMS/fetrends.htm.

8. John Heywood et al, "On the Road in 2035: Reducing Transportation's Petroleum Consumption and GHG Emissions," Laboratory for Energy and the Environment, Massachusetts Institute of Technology, July 2008, p. 122, http://web.mit.edu/sloan-auto-lab/research/beforeh2/otr2035/On%20the%20Road%20in%20 2035_MIT_July%202008.pdf.

One cannot overestimate the effectiveness of efficiency as a policy strategy for reducing oil demand and reducing emissions. In addition to high fuel economy requirements, policies to encourage new technology adoption and faster fleet turnover would also have a significant impact on fuel consumption and should be included as a complementary component of efficiency standards. Only a small percentage of the energy from gasoline used in an ICE goes toward moving the vehicle and the passenger inside. There are a variety of design and usage factors that decrease the efficiency of a vehicle. Engine energy loss can originate from component friction and wear or through operation of additional systems (parasitic loss) such as air conditioning and heat. Non-engine energy losses occur from idling in traffic or while parked, wind resistance, braking, and tire resistance, and additional inefficiencies are a result of improper tire inflation, poor maintenance, and driving habits.[9] While individually, each of these contributes a small share of energy use, a combination of improvements can have significant impact on overall vehicle operation and fuel savings.

Engine Technology

There have been vast technological improvements to the ICE, which have improved vehicle efficiency, allowing carmakers to increase horsepower while maintaining fuel economy. Such advances as electronic engine control and computer technology have allowed six-, seven-, and even eight-speed transmission designs, maximizing the vehicles' high efficiency (high gear) range in comparison to the three- and four-speed transmissions of a decade ago. One such design, Chrysler's new version of its 1960s Hemi, uses computer technology to control cylinder action when the vehicle is cruising at highway speed, thus rationing fuel consumption. While this particular technology has been available previously, electronic systems can now smoothly and imperceptibly shift between four and eight cylinders.[10]

There are many other new technologies and engines advancements that can increase ICE efficiency and reduce emissions. Turbocharged engines can provide increased performance in smaller engines allowing for smaller car designs with lower total vehicle weight. New combustion technologies are being explored that reduce the production of emissions in the engine or capture them at the point of exhaust. There is also much research being done on capturing wasted energy from engine heat and braking and converting it back to mechanical energy for reuse.

Vehicle Mass and Weight

Vehicle mass and weight are important components of efficiency, as heavier cars need more energy to accelerate and maintain speed. There are many ways to decrease the mass of vehicles, including design changes to the frame, using smaller engines, lighter transmissions and/or lighter materials, and decreasing total vehicle size.[11] Reducing vehicle weight would not reduce the percentage of energy wasted in the engine, but the amount of total energy needed would be less to move a lighter car.

---

9. Kara Kockelman et al., *GHG Emissions Control Options: Opportunities for Conservation,* commissioned paper for the National Academy of Science, October 20, 2008, http://www.ce.utexas.edu/prof/kockelman/public_html/NAS_CarbonReductions.pdf.

10. Joseph B. White, "The Car of the Future," *Wall Street Journal,* July 25, 2005.

11. Wenzel and Ross, "Increasing the Fuel Economy and Safety of New Light-Duty Vehicles," 2.

One of the main concerns with lighter vehicles is passenger safety. Although heavier vehicles are generally safer, quality, construction, and design may be a larger factor than weight in safety.[12] Advances in materials technology and vehicle design have made it possible to make lighter cars without significant safety disadvantages.[13] New high-strength steels and composites are much lighter than traditional materials and can be used for engine components, accessories, in addition to the chassis itself. These advancements in material science and design mean that even larger vehicles can be made lighter without sacrificing size or performance. And while higher costs are also a barrier to the use of new materials, these could be offset by the greater fuel economy consumers would realize through lighter vehicles. Furthermore, as more of these materials are adopted, the unit costs will become competitive with conventional supplies.

Vehicle Maintenance

Even the best-designed vehicle will not meet its fuel economy potential if it is weighed down with unnecessary items (especially in the trunk) and/or poorly maintained. Tires are particularly important, as rolling resistance is a major source of energy loss in vehicle operation.[14] Many drivers do not regularly check tire pressure or know what the correct pressure is for their particular tire and vehicle; as a result, tire under inflation is estimated for nearly a quarter of all vehicles on the road today.[15] Proper tire inflation together with regular vehicle maintenance could increase U.S. vehicle fuel economy anywhere from 3 to15 percent.[16]

One of the more obvious and beneficial steps that can be taken to lower carbon emissions from passenger vehicles and reduce the amount of fuel required to operate them is to improve effi-

---

12. "Research claiming that lighter vehicles are inherently less safe than heavier vehicles is flawed, and [we conclude] that other aspects of vehicle design are more important to the on-road safety record of vehicles." See Wenzel and Ross, "Increasing the Fuel Economy and Safety of New Light-Duty Vehicles," abstract. See also Laura Schewel and Noah Buhayar, "Picking a Safer Car for You, Your Family, and the Planet," Yahoo! Green, http://green.yahoo.com/blog/amorylovins/27/picking-a-safer-car-for-you-your-family-and-the-planet.html; Research and Innovative Technology Administration, *National Transportation Statistics* (Washington, D.C.: Bureau of Transportation Statistics, 2008), table 2-20, http://www.bts.gov/publications/national_transportation_statistics/html/table_02_20.html; National Highway Traffic Safety Administration (NHTSA), "Alcohol-Related Fatalities and Alcohol Involvement Among Drivers and Motorcycle Operators in 2005," in *Traffic Safety Facts* (Washington, D.C.: U.S. Department of Transportation, 2005), http://www-nrd.nhtsa.dot.gov/Pubs/810644.PDF; NHTSA, "Alcohol-Impaired Driving," in *Traffic Safety Facts* (Washington, D.C.: U.S. Department of Transportation, 2006), http://www-nrd.nhtsa.dot.gov/Pubs/810801.PDF; NHTSA, "2006 Traffic Safety Annual Assessment—A Preview," in *Traffic Safety Facts* (Washington, D.C.: U.S. Department of Transportation, July 2007), http://www-nrd.nhtsa.dot.gov/Pubs/810791.PDF; NHTSA, *The Impact of Driver Inattention on Near-Crash/Crash Risk: An Analysis Using the 100-Car Naturalistic Driving Study Data* (Washington, D.C.: U.S. Department of Transportation, 2006), http://www-nrd.nhtsa.dot.gov/departments/nrd-13/810594/images/810594.pdf; and Ray Resendes, "Advanced Safety Research: HATCHI-NHTSA Research Exchange," March 1, 2006, http://www.nhtsa.dot.gov/staticfiles/DOT/NHTSA/NRD/Articles/Public%20Meetings/Presentations/HATCI/0306Resendes.pdf.
13. Amory Lovins, *Winning the Oil End Game* (Colorado: Rocky Mountain Institute, 2004).
14. Kara Kockelman et al., *GHG Emissions Control Options: Opportunities for Conservation.*
15. NHTSA, *Federal Motor Vehicle Safety Standards; Tire Pressure Monitoring Systems; Controls and Displays* (Washington, D.C.: U.S. Department of Transportation, 2004), http://www.nhtsa.dot.gov/cars/rules/rulings/TirePresFinal/TPMSfinalrule.pdf.
16. U.S. Department of Energy, "Gas Mileage Tips," http://www.fueleconomy.gov/feg/drive.shtml.

ciency. Proper maintenance, implementation of advanced engine technologies, decrease in vehicle weight, and stricter fuel economy standards are all important and available options to increase the efficiency of end-use energy.

New Technologies
Diesel

Advanced diesel engines dominate the European market and average 5 mpg more than the total EU fleet fuel economy, even given the greater carbon intensity of diesel fuel to gasoline. Diesel engines utilize the latest in advanced engine technology. Direct injection improves fuel economy and reduces vehicle emissions by ensuring that the fuel is combusted evenly in the engine at higher temperatures.

High-mileage diesel models are just being introduced into the United States. As with smaller and lighter vehicles, however, public perception and acceptance of diesel engines is the largest barrier to market penetration. While modern diesel engines are imperceptible to their gasoline counterparts and offer many benefits, the cost of diesel fuel (relative to gasoline) as well as older concerns related to engine noise and pollution may continue to retard their sales growth in the United States. Improving gasoline engines by adopting characteristics from the diesel engines, such as advanced timing controls and dynamically optimized injection systems, may be a more widespread option.

Flex Fuel

Vehicles that can run on gasoline or high-content alternative fuel blends are characterized as flexible-fuel vehicles (FFVs or "flex-fuel"). This technology is fully commercial, in particular the ethanol FFV, which is capable of running on E85 (a fuel blend of 15 percent gasoline to 85 percent ethanol). There are more than 7 million FFVs in the United States today.[17] Many Americans are unaware that the vehicle they drive is a FFV.[18] For large-engine vehicles, such as heavy SUVs, the flex-fuel designation offers advantages that are more significant to the manufacturer with respect to fleet mileage ratings than to improved fleet efficiency and lower total emissions.[19]

Many believe that ethanol specifically, rather than biofuels more broadly, is a practical next step in energy diversification and independence. But one dilemma facing FFVs is the lack of E85 or other gasoline blend fueling stations. There are currently more than 120,000 retail gasoline stations in the United States, and less than 1 percent have E85 fuel pumps.[20] As only a small percent-

17. National Renewable Energy Laboratory (NREL), "Light Duty E85 FFVs in Use (1998–2008)," http://www.afdc.energy.gov/afdc/data/vehicles.html.

18. Jeffrey Goettemoeller and Adrian Goettemoeller, *Sustainable Ethanol: Biofuels, Biorefineries, Cellulosic Biomass, Flex-Fuel Vehicles, and Sustainable Farming for Energy Independence* (Maryville, MO: Prairie Oak, 2007).

19. Alexei Barrionuevo and Micheline Maynard, "Dual-Fuel Vehicles Open Mileage Loophole for Carmakers," *New York Times*, August 31, 2008.

20. According to the 2002 U.S. Census Bureau, there were 121,446 gasoline stations in the United States. As of 2008, it was estimated that there were 1,900 stations with E85. U.S. Census Bureau, http://quarterhorse.dsd.census.gov/TheDataWeb_HotReport/servlet/HotReportEngineServlet?emailname=bh@boc&filename=sal1.hrml&20071127090603.Var.NAICS2002=447&forward=20071127090603.Var.NAICS2002; http://e85vehicles.com/e85-stations.htm.

age of the vehicles are flex-fuel (E85 capable), there is a business model question as to why retail gas station owners would agree to install E85 pumps unless a significant portion of their customer base owned E85 vehicles. Installation of E85 pumps makes inherent good sense in areas where there is either a high concentration of flex-fuel vehicles and/or a close proximity to ethanol blenders, but is of questionable economics for the vast majority of retail outlets. As discussed previously, E85 distribution requires separate blending and storage facilities, and there is considerable debate as to who should bear the costs of these infrastructure upgrades if E85 were to become more widely available and/or mandated across the United States.

At the time of this writing, consideration is under way for raising the national ethanol blend cap from 10 percent to 12 or 15 percent, although auto manufacturers' warranty coverage may limit that adjustment pending further testing.

HEVs and PHEVs

Hybrid electric vehicles (HEVs) have an internal combustion engine as well as an electric motor powered by a battery. Depending on the design, either engine can power HEVs independently. Regenerative braking and batteries provide the ability to recapture and store energy, while the gasoline engine is shut off while idle. Hybridization offers material advantages for vehicle mileage efficiency. Depending on the size and configuration of the engine, HEVs are able to obtain twice the gas mileage of the average mid-sized car sold in the United States, saving 350 gallons of gasoline per year per vehicle. As with the standard ICE, improvements in HEV technology can be used for greater fuel economy or for greater engine performance, largely dependent on consumer preference.

The plug-in hybrid electric vehicle (PHEV) is a transformative technology that can effectively change the transportation fuel landscape. Recent advancements in battery technology (e.g., shorter charge time, greater storage capacity to allow for increased mileage and distance, etc.) hold the promise of doubling or tripling fuel economy for short-haul trips. This technology creates an opportunity where the consumer has a choice of electric power from a variety of sources and liquid fuels. As such, PHEVs offer the consumer an arbitrage opportunity, much like flex-fuel vehicles do for gasoline versus E85, to choose the least-cost, most-convenient fuel to power their vehicle.

Initial studies point to the possible synergies of low carbon electricity generation and PHEVs, as integration with the electric grid can be used for both charging the vehicle and for serving as a mobile storage capability. At a national level, if PHEVs were allowed to provide power to the grid, so-called V2G mode, renewable electricity generation capacity could double.[21] There are, however, carbon concerns as well. Studies show that overall emissions depend on what time of day the vehicle is charged and what the source of electricity is. Vehicles charged during evening peak hours may increase emissions compared to a standard ICE.[22]

The greatest challenges for HEVs and PHEVs are the increased costs related to battery development. Advanced manufacturing is underdeveloped in the United States and contributes to the

21. See NREL, "Plug-In Hybrid Electric Vehicles and Wind Energy," http://www.nrel.gov/analysis/winds/pdfs/wind_phev_poster.pdf.

22. See P. Denholm and W. Short, *An Evaluation of Utility System Impacts and Benefits of Optimally Dispatched Plug-In Hybrid Electric Vehicles* (Golden, CO.: NREL, October 2006), http://www.nrel.gov/docs/fy07osti/40293.pdf.

high cost of new vehicles. There is a lack of domestic manufacturing capability, and a critical core skill set in battery technologies and production must be scaled up for prices to come down. Cost challenges are determined by a number of factors, but primarily driven by battery capacity, size, and weight. Although there is a strong incentive to increase battery charge rates by at least two orders of magnitude, increase energy density, and decrease volume and weight—all while maintaining life-cycle capability, safety, and manufacturability—inconsistent policies and different global standards create confusion and thwart investment.

An increased number of PHEVs would create higher demand for electricity with various economic, environmental, and security implications. A full well-to-wheels power analysis would depend on the electric power source and the efficiency of the grid. Before advocating full electrification of the transportation system, policymakers will have to thoughtfully assess the fuel sources that are expected to provide the electric power. Coal provides fully half of current power generation in the United States, but increasing concerns over climate change regulation have stalled the construction of new coal-fired power plants. Advocates of nuclear energy frequently point to its environmental advantages, but scale-up challenges remain, as do concerns related to waste disposal, security, and proliferation. Natural gas, while cleaner than coal, comes with its own set of medium- to longer-term supply concerns. Renewables offer obvious advantages, but intermittency issues raise questions of reliability for purposes of meeting load demands 24 hours a day, 7 days a week, 365 days a year.

Fuel Cells

Fuel cells are another key enabling technology for HEVs and PHEVs. Fuel cells are most frequently discussed within the context of hydrogen as fuel, in which hydrogen and oxygen are catalytically recombined in the fuel cell to create electricity and water. However, fuel cells can also use other fuels such as methanol or organic materials (for microbial fuel cells). While fuel cells have proven robust in high-value-added systems, such as space vehicles and select larger-scale stationary power systems, they have yet to be proven for transportation except in early demonstration programs. High costs, power capacity, and fuel availability/storage and safety are among the primary issues facing transportation applications. Although fuel cells may offer great promise for transportation, much more is required in the areas of research, development, and demonstration. A more detailed analysis of electric vehicle options requires a complimentary, in-depth analysis of the reliability and efficiency of the electric delivery system, accessibility for recharge opportunities, and the security and environmental considerations of the various choices of fuel that will be used for power generation.

Just as with alternative fuels, policies aimed at encouraging specific new vehicle technologies, while highly desirable, must be met with cautious optimism. It took decades for the transportation infrastructure to evolve, and the risks and benefits of new pathways must be carefully examined before one is picked. Setting consistent policy goals through a vehicle performance standard is generally more effective than picking technology "winners." Transformation of the vehicle infrastructure will take coordinated efforts that account for fuel supply availability, consumer preferences, manufacturing capabilities, and a host of other sectors. What is needed is a commitment to innovation and common goals, while producing products that Americans want to drive, using scalable and sustainable business practices.

4 | POLICY OUTLOOK

Policy Outlook Is Increasingly Complex

In the past, energy policymakers had the luxury of responding to a single driver, such as supply security or economic growth. Today, we are facing a more complex paradigm in which a number of major concerns (whose remedies are often competing or contradictory) have to be considered in tandem when attempting to formulate effective policy decisions. These elements include energy security, reliability, economic prosperity, environmental sustainability, technology, infrastructure, investment needs, and the realistic time frames needed to implement the changes to such a large-scale system.

The world is currently at an energy crossroads and in need of U.S. leadership. The development of reliable, cost-effective, and environmentally sensitive energy is one way that the United States can maintain and strengthen its role as a world leader. While there is currently a great deal of debate and discussion on a variety of smaller issues, it is prudent to step back to view the larger picture. Effective and productive policies to address all of our economic, environmental, and energy security needs are possible, but only if policymakers have a holistic understanding of the complexities of the issues and are willing to work across committees and collectively embrace comprehensive solutions. Short-term, unsustainable solutions that fail to address these underlying concerns will waste valuable time and resources, ultimately failing to meet collective needs.

While U.S. energy intensity (energy per GDP) has continued to decrease as the economy grows and transitions toward both more efficient use of energy and less energy-intensive industries, overall energy demand in absolute terms, under business-as-usual assumptions, continues to grow. U.S. energy expenditures as a percentage of GDP have declined in real terms for the past 25 years—until about 2000. Energy expenditures constitute roughly 10 percent of GDP. The United States is at the end of the era of "cheap" energy.

Limited experience in dealing with gasoline prices in excess of $4.00 per gallon make it difficult to conclude how longer-term consumption behavior (given new technology, transit, and work options) will be affected by volatile prices going forward. The current economic downturn has had a devastating effect on auto manufacturers, but even before the financial crisis hit, persistently high prices clearly impacted gasoline sales in the United States, with consumption year over year down 4 percent from 2007 to 2008. Consumer behavior changed: from July 2007 to July 2008, car sales increased by 0.3 percent and light truck sales fell 25.2 percent,[1] and according to the American Public Transportation Association, trips on public transportation increased 3.4 percent in the first quarter of 2008 compared with the first quarter of 2007.[2] Increased energy costs, food

---

1. Nick Bunkley, "U.S. Vehicle Sales Fall 13.2% Amid High Gas Prices," *New York Times*, August 2, 2008.

2. American Public Transportation Association, "Transit Ridership Report," June 27, 2008, http://www.apta.com/research/stats/ridership/riderep/documents/08q1cvr.pdf.

costs, reduction in household assets, and tightened consumer credit emerged as critical issues for the U.S. economy as a whole and a central focus of the 2008 presidential campaign. Immediately following President Barack Obama's inauguration, the White House energy and environment Web page identified oil import reduction, improved energy efficiency standards, putting 1 million high fuel efficiency vehicles on the road by 2015, and significantly reducing GHG emissions as top priorities for the new administration. At issue, however, is the ability and mechanisms chosen by the new administration.

Carbon Regulation Is the Game Changer

Clearly, the biggest "game changer" on the energy landscape today is the regulation of CO_2 emissions. Over the last several years, states across the nation proposed and in some cases passed measures to regulate and reduce CO_2 and other greenhouse gas emissions. The most notable progress made to date is by the Regional Greenhouse Gas Initiative (RGGI), a grouping of 10 northeast states with a functioning regional cap and trade program. Other states, particularly in the midwest, are poised to follow suit. This state-level action has bolstered support for similar policies at the federal level. In the last Congress there was a considerable increase in the number of proposals calling for the regulation of greenhouse gas emissions, the vast majority of which propose a cap and trade program. While a cap and trade program was unlikely to pass the last congress and almost certainly would have been vetoed by then-President Bush, it is a key pillar of the Obama administration's legislative agenda. A cap and trade proposal put forward by Representatives Henry Waxman and Edward Markey recently passed the House by a narrow margin but faces a tough battle to get through the Senate where political support will be harder to build. At this junction, its legislative fate remains uncertain even though congressional leaders have endorsed the notion of pursuing climate legislation during this Congress.

Of particular significance for the transportation market, the Supreme Court has ruled that the EPA has jurisdictional authority over mobile greenhouse gas emissions. The EPA is currently developing proposed rules to address this. Carbon emissions of the transportation fleet are one of the main policy issues actively under review by the EPA. Congress has multiple legislative proposals under consideration, some of which would regulate mobile sources of greenhouse gas emissions, building off the recently enacted increase in CAFE standards. The patchwork of energy and climate policy that has emerged has created a new legal storm over jurisdiction of emissions regulation.

Driven by high oil prices, concerns over increasing reliance on imported oil, climate change concerns, and the potential for rural development in the Midwest, policymakers enacted a renewable fuels standard (RFS) as part of the *Energy Policy Act of 2005* (EPACT 2005) as a mandatory increase in the amount of renewable fuels (mostly biofuels) in the transportation fuel mix. The combination of the mandate and high oil price environment, in addition to the aforementioned 51 cent per gallon blending credit and tariff protection that effectively blocked low-cost import competition, produced pronounced growth in the biofuels industry. The U.S. ethanol industry produced 9 billion gallons of the biofuel in 2008, with production capacity estimated to be 10.6 billion gallons in 2009. As of January 2009, 26 plants with 2 billion gallons per year production capacity were under construction, but market conditions may materially impact construction schedules and force a number of plant closings.[3]

As part of the *Energy Independence and Security Act of 2007* (EISA 2007), signed into law on December 19, 2007, the 110th Congress, starting in 2008, raised each annual RFS target beyond the volumes set forth in EPACT 2005 and therefore expanded the RFS to an ultimate goal of 36 billion gallons per year by 2022. In an attempt to address some of the environmental and food security issues of conventional biofuels, EISA 2007 mandates that, starting in 2009, a certain portion of the RFS target must be met by "advanced biofuels" (eventually reaching 21 billion gallons out of the 36 billion gallons required by 2022) and that all renewable fuels must yield at least a 20 percent reduction in greenhouse gas emissions compared to conventional gasoline or diesel.[4]

As previously discussed, increased use of ethanol in general and E85 in particular is the subject of considerable debate. Recent studies that examine the life-cycle carbon impacts of various ethanol production pathways have come to various conclusions; these studies and their assumptions, as well as the underlying science, deserve careful scrutiny. Full life-cycle impact analyses are challenging and depend heavily on the scenario of study. Avoiding immediate reaction and mindfully calling for careful science to inform policymakers is crucial, especially given the production goals of recent policies in the United States. Other considerations, such as infrastructure needs and costs, international trade, food versus fuel, vehicle efficiency/technology, consumer preferences, and other policies to reduce vehicular mileage traveled and fuel consumption will need to be carefully considered.

One of the major elements of EISA 2007 was to increase U.S. vehicle mileage standards. EISA sets out new vehicle efficiency guidelines that are expected to raise fuel economy standards by 40 percent and reduce U.S. demand for oil.[5] The new law raises fuel economy standards for cars and lights trucks to 35 miles per gallon by model year 2020. Additionally, medium- and heavy-duty trucks will have to comply with a new fuel economy program.[6] In May 2009, President Obama unveiled new regulations that would accelerate greater fuel economy, requiring an average standard of 35.5 mpg by 2016.[7] On an international level, however, even the new U.S. fuel economy standards remain among the lowest of developed nations. The framework for new policy development is increasingly complex, spanning local, regional, national, and global issues. Energy and transportation are two elements of the broader policy setting. Trade-offs will be required, especially give the state of the economy and the complexity of issues ranging from farm income, food prices, government research development and deployment (RD&D) funding, fuels regulations, fuel economy regulations, employment, trade agreements, security concerns, and last but not least climate negotiations.

---

3. Renewable Fuels Association, "Ethanol Industry Statistics," http://www.ethanolrfa.org/industry/statistics/.

4. *Energy Independence and Security Act of 2007*, Public Law 110-140, 110th Cong., 2nd sess. (December 19, 2007), http://www.govtrack.us/congress/bill.xpd?bill=h110-6.

5. White House, "Fact Sheet: Energy Independence and Security Act of 2007," press release, December 19, 2007, http://www.whitehouse.gov/news/releases/2007/12/20071219-1.html.

6. Fred Sissine, *Energy Independence and Security Act of 2007: A Summary of Major Provisions* (Washington, D.C.: Congressional Research Service, December 21, 2007), http://energy.senate.gov/public/_files/RL342941.pdf.

7. President Barack Obama, memorandum to the secretary of transportation, January 26, 2009, http://www.whitehouse.gov/the_press_office/Presidential_Memorandum_Fuel_Economy.

A New Framework Is Needed

The framework for creating policy, if structural change is to occur in the industry and the economy at large, will be most effective if it addresses the national public policy goals of energy security, environmental stewardship, and economic prosperity, while recognizing the complex global and sector-specific (e.g., agriculture, food, fuel, international trade, water) interdependencies. Policies that create a long-term, stable investment environment, while avoiding the picking of technology winners (and losers) or molecule-specific mandates, and that do not (mis)lead to large-scale stranded investments, will encourage and enable investment and innovation while reaching the public policy goals.

In the last 150 years, the United States has undergone at least two energy transitions. From 1850 to 1910, we moved from a "renewable energy" economy to one increasingly reliant on coal as our primary energy source. Between 1910 and 1970, we again transitioned from overwhelming reliance on coal to more than 60 percent reliance on oil and natural gas with an energy sector characterized by high-density resources, large "economy of scale" conversion in power plants and refineries, and low marginal cost transmission (pipelines, electric grid). Today, the question before us is which "pathway" we pursue, given the complex challenges of energy policy and other societal goals and national objectives. We are now confronting a third transformation—one that requires us to recalibrate our policies in light of changing landscapes and growing concerns associated with energy reliability and security, economic well-being, and environmental stewardship of our planet.

EPILOGUE
Erik R. Peterson[1]

Despite the recent downturn in prices occasioned, at least in part, by the global recession, most analysts agree that the longer-term forecast for global energy demand meeting projected growth is simply unsustainable. At some point, as economies recover, driven in large part by continued and rapid economic growth in emerging economies such as China and India, the risks and challenges associated with expansion of conventional sources of energy suggest the need to accelerate development of all available energy sources—conventional and unconventional, nuclear and renewables—along with a healthy dose of technology-driven advances in efficiency. Each source of energy carries with it formidable uncertainties—including but not limited to geopolitical risks, regulatory actions, economic changes, financial shifts, and innovation and diffusion of relevant technologies. No single energy supply pathway can satisfy the future demand levels that are currently forecast even with expanded supply, and there are profound uncertainties about whether a stable and secure energy environment can be achieved.

Rising energy demand will also serve to punctuate the profound challenges associated with balancing long-term economic, security, and environmental goals. There can be little doubt that actions taken to address global climate change will alter the terms of reference for the debate on energy. It will likely affect demand growth and the underlying energy economics and influence national policies among all nations. The trade-offs between traditional notions of energy security and the new environmental realities—carbon management in particular—will present leaders with decisions ever more complex and challenging.

Recent events reflect how profoundly far reaching the challenge of energy has already become. Early last year, rising energy prices were identified as one of the key reasons why in societies across the world, agricultural dislocations precipitated food riots, increasing pressure on the poor, slowing the pace of economic development, and impeding economic growth more generally. According to the UN World Food Program, these forces together created a "silent tsunami" affecting more than 100 million across the world.[2] While forces not relating to energy are at work, including bad weather, lack of investment, export restrictions, and other trade-distorting practices, the situation was exacerbated by significantly increased energy prices and the diversion of land and agricultural production to fuel production.

Added to the equation is yet another critical dimension—that of water. Expanding energy production to satisfy projected global demand will require even greater amounts of water. Such diversions will put additional pressure on agriculture, which accounts for an estimated 70 percent of the demand on current freshwater resources. It will also affect the 1.1 billion persons who do

---

1. Erik R. Peterson is senior vice president, director of the Global Strategy Institute, and holder of the William A. Schreyer Chair in Global Analysis at CSIS.

2. See "VNR: The Silent Tsunami," United Nations World Food Program, http://www.wfp.org/english/?ModuleID=148&Key=207.

not have access to safe drinking water and the 2.4 billion who have inadequate sanitation.

This energy-food-water triangle—and the shifting dynamics among the interrelated corners—suggests the potential for more onerous constraints when it comes to resource management in the future. It also suggests the potential for mounting resource nationalism as the dislocations in these interrelated areas become more pronounced.

For these reasons and others, the case for developing a more robust understanding of the many complexities associated with the broad domain that we can call "alternative energy" is becoming more compelling by the day. It goes beyond assessing the state of alternative energy as a response to volatilities in petroleum markets, as profound as they have been over recent years. When increases in population (by some 2.5 billion to a total of 9.2 billion through the year 2050), relentless aggregate energy demand growth, complex linkages between energy and water and agriculture, and growing impacts of environmental degradation (global warming, in particular) are all taken into account, it becomes essential that we develop a clearer understanding of the "alternative energy" variable in the global energy equation.

Such an assessment defies the bumper-sticker solutions and sound-bite positions that various observers might and do advocate, especially at this time of pronounced energy market volatility and high prices at the pump. The trade-offs are tremendously complicated, the complexities are more and more evident, the diverse technologies at work are all in flux, and the policy environments are extremely fluid. Any assessment of the state of alternative energy is a mere snapshot, at a single moment in time, of where things are and where they need to go. The outlook and the contours of the debate that will influence the alternative energy and broader energy trajectories depend on myriad elements—political, social, economic, regulatory, financial, environmental, technological, and security—that are changing constantly.

All this suggests tremendous challenges to policymaking, which across the world has been marked by more reactive than proactive positions and policies, by tactics rather than strategy, and by segmented rather than interdisciplinary analysis. What should be clear to leaders, however, is that they are on countdown—that they cannot continue to prevaricate as the energy (and resource) dislocations around them become even more pronounced.

CONFERENCE PARTICIPANTS

Robert Anex
Associate Professor
Department of Agricultural and Biosystems
    Engineering
Iowa State University

Dan Arvizu
Director
National Renewable Energy Laboratory (NREL)

Charles H. Bank
President
R.L. Banks & Associates Inc.

Carl Bauer
Director
National Energy Technology Laboratory (NETL)
U.S. Department of Energy

Eldon Boes
Director, Energy Analysis Office
NREL

Michael Brylawski
Vice President of Corporate Strategy
Bright Automotive

Vladimir Bulovic
*Associate Professor of Electrical Engineering and
    Computer Science*
MIT

Lou Burke
Manager of Emerging Technology
ConocoPhillips

Gerbrand Ceder
*R.P. Simmons Professor of Materials Science and
    Engineering*
MIT

Terry Cullum
*Director of Corporate Responsibility and Environment
    and Energy*
General Motors Corp.

Reid Detchon
Executive Director
Energy Future Coalition

Charles T. Drevna
Director of Technical Advocacy
National Petrochemical and Refiners Association

Bruce Everett
Edmund A. Walsh School of Foreign Service
Georgetown University

Bill Frey
Global Business Director
Biofuels Business DuPont

David Garman
*Assistant Secretary for Energy Efficiency and
    Renewable Energy (2001–2008)*
U.S. Department of Energy

Henry (Hank) Habicht
Vice Chairman
Global Environment and Technology Foundation

Howard Herzog
Principal Research Engineer
Laboratory for Energy and the Environment

John Heywood
*Director and Sun Jae Professor of Mechanical
    Engineering*
Sloan Automotive Laboratory
MIT

Note: These individuals participated in the CSIS Alternative Transportation Fuels and Vehicle Technologies series. The information in the report, except where noted, is based on the information given in their presentations and from the conference discussions. The authors gratefully acknowledge their many important insights and their participation in the series.

Susan Hockfield
President
MIT

Yang Shao Horn
Assistant Professor
Department of Mechanical Engineering
MIT

Mujid Kazimi
Professor of Nuclear and Mechanical Engineering
MIT

Elizabeth A. Lowery
Vice President of Environment and Energy
General Motors Corp.

Ernest J. Moniz
*Cecil and Ida Green Professor of Physics and
    Engineering Systems; Codirector of the Laboratory
    for Energy and the Environment*
MIT

Shirley Neff

Tad Patzek
Professor of Geoengineering
University of California–Berkeley

Donald Paul
Executive Director
Energy Institute
University of Southern California

Steve Plotkin
Transportation Energy Analyst
Argonne National Laboratory

John Reese
Fuels Product Management Adviser
U.S. Shell Oil Products

Ken Roberts
Senior Vice President of Business Development
Syntroleum

John Sheehan
Senior Strategic Analyst
NREL

Adam J. Schubert
U.S. Strategy Manager
BP Biofuels

Robert Williams
Senior Research Scientist
Princeton Environmental Institute

Robert Wimmer
National Manager
Toyota Motors North America

GLOSSARY

A

Advanced timing control—a system that reduces the amount of unburned fuel exhausted during the startup operation of an engine, maximizing its efficiency

Anaerobic gasification—breakdown of hydrocarbons into a syngas by carefully controlling the amount of oxygen present, e.g., the conversion of coal into gas

B

Biobutanol—butanol from biomass, can be used as fuel in an internal combustion engine

Biodiesel—nonpetroleum-based diesel fuel consisting of short-chain alkyl (methyl or ethyl) esters, typically made by transesterification of vegetable oils or animal fats, which can be used—alone or blended with conventional petrodiesel—in unmodified diesel-engine vehicles

Biomass—living and recently dead biological material that can be used as fuel or for industrial production

Butanol—primary alcohol with a four-carbon structure and the molecular formula of $C_4H_{10}O$; belongs to the higher alcohols and branched-chain alcohols

C

CAFE standards—Corporate Average Fuel Economy standards are federal regulations intended to improve the average fuel economy of cars and light trucks (trucks, vans, and sport utility vehicles) sold in the United States

Carbon mitigation strategies—broad term for the host of strategies aimed at reducing carbon emissions including low emissions technology, underground carbon storage, incentives to reduce carbon emissions, use of alternative low carbon-emitting fuels, and others

Carbon-to-liquids—fuels converted from nonpetroleum carbon sources such as coal, oil shale, and biomass feedstocks into usable liquid fuels

Carbonate matrix—the fine-grained mass of a carbonate material in which larger grains or crystals are embedded

Cassava—also known as manioc, casava, or yuca (Manihot esculenta); a woody shrub of the Euphorbiaceae (spurge family) native to South America that is extensively cultivated as an annual crop in tropical and subtropical regions for its edible starchy tuberous root, a major source of carbohydrates

Cellulosic ethanol—a type of biofuel produced from lignocellulose, a structural material that comprises much of the mass of plants

Cellulosic processing—method in which the long chains of sugar molecules are broken down to free the sugar, before it is fermented for alcohol production; there are two major forms of cellulose hydrolysis (cellulolysis) processing: a chemical reaction using acids (chemical hydrolysis) or an enzymatic reaction

Chassis—the supporting frame of a structure (as an automobile or television); the frame and working parts (as of an automobile or electronic device) exclusive of the body or housing

Coal-to-liquids (CTL)—process that converts coal into usable liquid fuel form through liquefaction processes

D

Densification—to increase in density

Dimethyl ether (DME)—a colorless, gaseous, water-soluble ether that has the formula CH_3OCH_3 or as its empirical formula, C_2H_6O (which it shares with ethanol); a clean-burning alternative to liquified petroleum gas, liquified natural gas, diesel and gasoline; can be made from natural gas, coal, or biomass

Dynamically optimized injection system—system that determines fuel demands of an engine based on engine speed and power produced by the engine at a given time, in order to optimize the engine's power output for a load while reducing engine emissions

E

Esterification—chemical reaction in which an alcohol and an acid react to produce an ester; employed in the production of biofuels

F

Feedstock—raw material required for an industrial process

Fischer-Tropsch process—catalyzed chemical reaction in which carbon monoxide and hydrogen are converted into liquid hydrocarbons of various forms; Fischer-Tropsch liquids are these products

Flex-fuel vehicle—vehicle that can operate on: 1) alternative fuels (such as E85); 2) 100 percent petroleum-based fuels; or 3) any mixture of an alternative fuel (or fuels) and a petroleum-based fuel; flex-fuel vehicles have a single fuel system to handle both alternative and petroleum-based fuels

Fuel cells—device capable of generating an electrical current by converting the chemical energy of a fuel (e.g., hydrogen) directly into electrical energy

Fuel oxygenate—chemical containing oxygen that is added to fuels, especially gasoline, to make them burn more efficiently

G

Gasification—method for converting coal, petroleum, biomass, wastes, or other carbon-containing materials into a gas that can be burned to generate power or processed into chemicals and fuels

Gas-to-liquids (GTL)—process that combines the carbon and hydrogen elements in natural gas molecules to make synthetic liquid petroleum products, such as diesel fuel

Glycerine—a liquid by-product of biodiesel production; used in the manufacture of dynamite, cosmetics, liquid soaps, inks, and lubricants

H

Hybrid electric vehicle (HEV)—combines the internal combustion engine of a conventional vehicle with the battery and electric motor of an electric vehicle; it offers low emissions, with the power and range of conventional vehicles

Hydrocracking—a refining process that uses hydrogen and catalysts with relatively low temperatures and high pressures for converting middle-boiling or residual material to high-octane gasoline, reformer charge stock, jet fuel, and/or high-grade fuel oil

Hydrogenation—treatment of substances with hydrogen and suitable catalysts at high temperature and pressure to saturate double bonds

Hydrothermal liquefaction—the reaction of solid biomass and water at elevated temperature and pressure in the presence of a catalyst; the main product is biocrude

Hydrous/wet biomass—animal manure, biosludge, residuals from fermentation facilities, and food garbage

J

Jatropha—genus of approximately 175 plants, shrubs, and trees native to Central America; oil from jatropha seeds can be used in a standard diesel engine, while the residue can be processed into biomass; cited as one of the best candidates for future production of biodiesel

L

Lignocellulose—substance constituting the essential part of woody cell walls and consisting of cellulose closely associated with lignin; fermentation of lignocellulosic biomass to ethanol can provide energy feedstock

M

Miscanthus—genus of approximately 15 species of perennial grasses native to subtropical and tropical regions of Africa and southern Asia; bioenergy crop

P

Petroleum coking—petroleum coke (often abbreviated petcoke) is a carbonaceous solid derived from oil refinery coker units or other cracking processes

Plug-in hybrid vehicle (PHEV)—hybrid vehicle with batteries that can be recharged by connecting a plug to an electric power source

Polygeneration—energy supply system, which delivers more than one form of energy to the final user, for example: electricity, heating, and cooling can be delivered from one polygeneration plant

Pyrolysis—the chemical decomposition of organic materials by heating in the absence of oxygen or any other reagents, except possibly steam

R

Rapeseed—bright yellow flowering member of the mustard/cabbage family; rapeseed oil is used in the manufacture of biodiesel

Refined petroleum products—derived from crude oils through processes such as catalytic cracking and fractional distillation; examples of refined petroleum products include gasoline and kerosene

Regenerative braking—mechanism that reduces vehicle speed by converting some of its kinetic energy into another useful form of energy

Renewable energy—effectively uses natural resources such as sunlight, wind, rain, tides, and geothermal heat, which are naturally replenished; renewable energy technologies range from solar power, wind power, hydroelectricity/micro hydro, biomass, and biofuels for transportation

Retorting—process for the recovery of volatile material from solid carbonaceous material, such as oil shale, tar sands, coal, and lignite, which requires controlling the atmosphere surrounding the carbonaceous material during the process

S

Sorghum—genus of numerous species of grass; used as feedstock in the production of biobutanol

Strategic Petroleum Reserve (SPR)—petroleum stocks maintained by the U.S. government for use during periods of major supply interruption

Switchgrass—warm-season grass found in the central North American tallgrass prairie; bioenergy crop

Synfuel—synthetic fuel; any liquid fuel obtained from coal, natural gas, or biomass

X

X Prize—competition held by the X Prize Foundation meant to bring about solutions to major global challenges; for the third X Prize, teams will compete for multimillion dollar cash prizes by designing and building super-efficient vehicles that can achieve 100 miles per gallon

ABOUT THE AUTHORS

Douglas Arent is director of the Strategic Energy Analysis Center at the National Renewable Energy Laboratory (NREL). He specializes in strategic planning and financial analysis competencies; clean energy technologies and energy and water issues; and international and governmental policies. In addition to his NREL responsibilities, Arent is a member of the U.S. government review panel for Intergovernmental Panel on Climate Change (IPCC) reports and is a visiting fellow at CSIS. Arent was appointed in 2008 to serve on the National Academy of Sciences panel on "Limiting the Magnitude of Future Climate Change." Arent is on the Executive Council of the U.S. Association of Energy Economists and is on the Advisory Board of E + Co, a public purpose investment company that supports sustainable development across the globe. He serves on the Chancellor's Committee on Energy, Environment, and Sustainability Carbon Neutrality at the University of Colorado. Arent was the chair of the Quantitative Work Group in support of the Clean and Diversified Energy Advisory Council of the Western Governor's Association. Prior to coming to NREL, he was a management consultant to clean energy companies, providing strategy, development, and market counsel. Previous positions held include: director of strategic marketing and business development at Network Photonics; director of Media Gateway Products and strategic planning manager at Lucent Technologies (now Avaya); and vice president of business development for Amonix, Inc. Dr. Arent has a Ph.D. from Princeton University, an M.B.A from Regis University, and a bachelor's of science from Harvey Mudd College in California.

Frank Verrastro is senior fellow and director of the CSIS Energy and National Security Program. His energy-related experience includes more than 25 years in energy policy and project management positions in both the U.S. government and the private sector. His government service includes staff positions in the White House (Energy Policy and Planning Staff) and the Departments of Interior and Energy, including serving as deputy assistant secretary for international energy resources. In the private sector, Verrastro served as director of refinery policy and crude oil planning for TOSCO (formerly the nation's largest independent refiner) and more recently as senior vice president for Pennzoil. His responsibilities at Pennzoil included government relations, both domestic and international; corporate planning; international risk assessment; and negotiations. He also served on the company's management and operating committees, as well as on the Environmental Safety and Health Leadership Council.

Verrastro holds a B.S. in biology/chemistry from Fairfield University and a master's degree from Harvard University, and he has completed the executive management program at the Yale School of Management. He has been an adjunct professor at the Elliott School of International Affairs at the George Washington University and at the University of Maryland and is a frequent presenter at the Foreign Policy Institute of the Department of State. He currently serves on the advisory board of the National Renewable Fuels Laboratory in Golden, Colorado.

Jennifer L. Bovair is program manager and research associate in the CSIS Energy and National Security Program, where she provides research and analysis on a range of energy issues, focusing on energy security, geopolitics of energy in Eurasia, alternative fuels and vehicle technologies, climate change and sustainability, in addition to managing program conferences and meetings. Bovair served as assistant to the leader of the geopolitics and policy groups on the landmark 2007 study by the National Petroleum Council for the U.S. secretary of energy. Before joining CSIS, she worked at the U.S. Department of Commerce's International Trade Administration in the Office of Energy and Environment, the U.S.-Russia Business Council, and the U.S. Civilian Research and Development Foundation. She received her M.A. from the Walsh School of Foreign Service at Georgetown University, with an honors certificate in international business diplomacy, and holds a B.A. from the University of Michigan.

Erik R. Peterson is senior vice president at CSIS, where he is director of the Global Strategy Institute. As director, he heads the Seven Revolutions Initiative, an internationally recognized effort to identify and forecast global trends out to the year 2025 and beyond. Peterson also holds the William A. Schreyer Chair in Global Analysis at CSIS. For his contributions to the Center, he received the 2006–2007 CSIS Trustees Award. Before joining CSIS, he was director of research at Kissinger Associates. Peterson serves on several advisory boards, including those of the X Prize Foundation, the Center for Global Business Studies at Pennsylvania State University, and the Center for the Study of the Presidency. He was a fellow of the World Economic Forum and has served on its Global Risk Network. He recently contributed a chapter entitled "Scanning the More Distant Future" to *For the Common Good: The Ethics of Leadership in the 21st Century* (Praeger, 2006). He has spoken before numerous groups and lectured in 14 countries. Peterson received his M.B.A. from the Wharton school, his M.A. from the Johns Hopkins University School of Advanced International Studies, and his B.A. from Colby College. He holds the Certificate of Eastern European Studies from the University of Fribourg in Switzerland and the Certificate in International Legal Studies from The Hague Academy of International Law in the Netherlands.